EVOLUTION'S WORKSHOP

Also by Edward J. Larson

Trial and Error:
The American Controversy over
Creation and Evolution

Sex, Race, and Science: Eugenics in the Deep South

Summer for the Gods: The Scopes Trial and
America's Continuing Debate over
Science and Religion

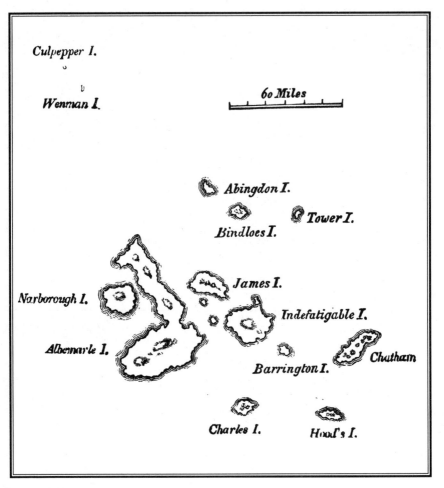

Chart of Galápagos Islands, from Charles Darwin, Journal of Researches, *2d ed. (1845)*
(Source: Used by permission of University of Georgia Libraries [UGA].)

EVOLUTION'S WORKSHOP

God and Science on
the Galápagos Islands

EDWARD J. LARSON

BASIC
BOOKS

A Member of the Perseus Books Group

Published by Basic Books,
A Member of the Perseus Books Group

Library of Congress Cataloging-in-Publication Data
Larson, Edward J. (Edward John)
    Evolution's workshop : God and science on the Galápagos Islands / Edward J. Larson.
       p. cm.
    Includes bibliographical references (p.).
    ISBN 0-465-03810-7
    1. Evolution (Biology)—Galápagos Islands.   2. Evolution.   I. Title.

QH366.2 .L375 2001
576.8'09866'5—DC 21

                                                                00-065159

Design by Elizabeth Lahey
FIRST EDITION

01 02 03 04 / 10 9 8 7 6 5 4 3 2 1

Dedicated to my teacher,
Ronald L. Numbers
and to my students

CONTENTS

PREFACE

MY WIFE AND I first visited the Galápagos Islands in 1992, on a side trip during our vacation in Ecuador. We were enchanted by the place. As a historian of science, I was especially intrigued by its past. I knew about Charles Darwin's legendary encounter with the islands' finches, of course, but there was obviously much more to the history of science on the Galápagos than simply Darwin. Our room overlooked Academy Bay, named for the California Academy of Sciences in far-off San Francisco. There must be a story behind that name—and another behind the bustling Charles Darwin Research Station down the road. There and then I resolved to investigate these matters further, hoping to cull enough material for an article or two. At least the effort might justify a return trip.

The deeper I looked, the more I found about the archipelago's rich history in science stretching back before Darwin. Even the Darwin story proved richer than its legend. Two years after I began my research, Jonathan Weiner published *Beak of the Finch*, his prizewinning book about Peter and Rosemary Grant's ongoing study of island birds. At the time, I was still back in the 1800s doing historical research but at once realized that Galápagos science has a vital future as well as a profound past. So I pushed on with research into the later 1800s and early 1900s, where I encountered the marvelous fieldwork sponsored by Walter Rothschild, the California Academy of Sciences, the American Museum of Natural History and Allan Hancock. Then came Julian Huxley, the Darwin Station and the emergence of ecology in Galápagos science following the Second World War. By this point, my project had evolved into a book.

Along the way, I incurred too many debts to ever fully acknowledge. For those of you who helped me, let me begin by expressing my gratitude. First, I thank those many institutions and individuals who have preserved material on the history of Galápagos science and shared it with me, most notably the Charles Darwin Research Station library and its librarian, Gayle Davis; the California Academy of Sciences and its archivists, Michele Wellck and Karren Elsbernd; the Museum of Natural History (London) and its archivist, Susan Snell; the American Museum of Natural History and its senior special collections librarian, Barbara Mathe; and the Allan Hancock Foundation at the University of Southern California, Scripps Institution of Oceanography, the Museum of Comparative Zoology at Harvard University, Cambridge University libraries, Stanford University archives and the Smithsonian Institution. I also benefited from information and insights from many individuals associated with Galápagos science and conservation, including Michele L. Aldrich, Robert Bensted-Smith, P. Dee Boersma, Robert I. Bowman, Annie Dillard, Peter R. Grant, Matthew J. James, Peter C. Lack, Alan E. Leviton, Craig G. MacFarland, Ernst Mayr, Godfrey Merlen, Marc L. Miller, Marc Patry, Duncan M. Porter, Patricia Robayo, Miriam Rothschild, Dolph Schluter, Vassiliki B. Smocovitis, Heidi M. Snell, Howard L. Snell, David A. Westbrook and John Woram. Special materials came from John N. Johnson relating to the critique of evolutionary science, and John J. Abele regarding the 1942 U.S. Naval Intelligence report on the archipelago. Finally, the University of Georgia libraries and their able staffs richly supplied my every need for published materials on Galápagos science. Except as otherwise indicated, all illustrations in this book come from out-of-copyright materials reproduced here by permission of the University of Georgia libraries.

My writing has also benefited greatly from the conceptual and editorial assistance of many friends. Throughout, William Frucht, my editor at Basic Books, labored to improve this book as much as I would let him. He has a particular gift for finding fat in the leanest of prose: he could have brought the Gettysburg Address in under two minutes. For the second time in a row, copy editor Michael Wilde trained his keen eye on my prose. There were others who helped in the writing process as well, especially Galápagos botanist and Darwin scholar Duncan M. Porter, who read every word of my bloated original manuscript; the incomparable Darwin biographer Janet Browne, who reviewed major portions of it; my

Athens, Georgia, neighbor, the gifted writer Philip L. Williams; my university colleague, Thomas M. Lessl; and the discerning Gerard McCauley. My thanks go to all of them.

Invaluable support came from other institutions and individuals as well. I gratefully acknowledge the University of Georgia, particularly my deans, Wyatt W. Anderson and David E. Shipley, and the Richard B. Russell Foundation, particularly board chair Charles C. Campbell, for their ongoing support of my work. Most of all, I love and appreciate my wife, Lucy, and our children, Sarah Marie and Luke Anders. During the course of my research and writing, they have suffered through my many absences and preoccupations without displaying the least jealousy toward giant tortoises and blue-footed boobies.

—E. L.

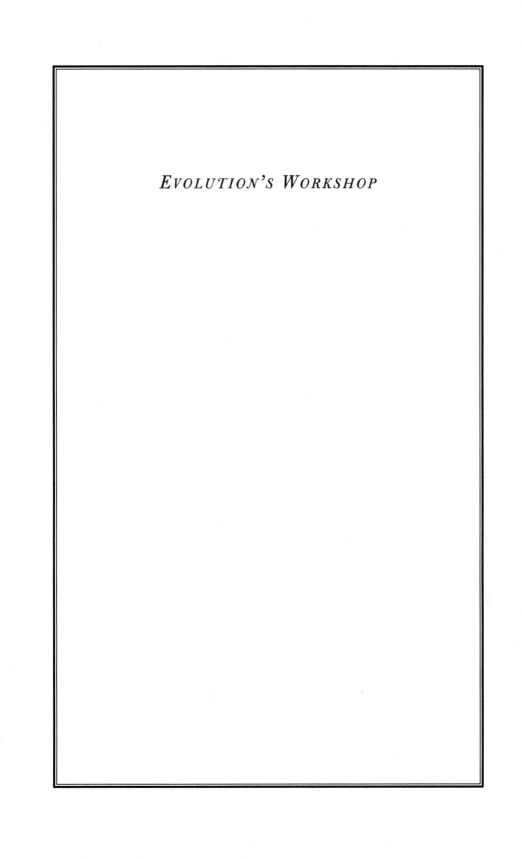

EVOLUTION'S WORKSHOP

ENCHANTMENTS
OLD AND NEW

IT WAS ON THE GALÁPAGOS in the early autumn of 1835 that Darwin took the first step out of the fairyland of creationism into the coherent and comprehensible world of modern biology," asserted the influential British biologist Julian Huxley, "for it was here that he became fully convinced that species are not immutable—in other words, that evolution is a fact." This, Huxley went on, "effected the greatest of all revolutions in human thought, greater than Einstein's or Freud's or even Newton's." Darwin's findings changed how scientifically informed people view nature, life and humankind. The central role played by the Galápagos Islands in this revolutionary moment eventually transformed that remote, desolate archipelago into a sacred site for science and a place of immense interest to biologists and eco-tourists alike. They came to call it "Darwin's Eden" and "The Galápagos Wonderland."[2]

People had not always seen the archipelago this way. At nearly the same time that a young and impressionable Charles Darwin first began to comprehend its marvels, another perceptive young visitor destined for greatness reflected a more commonly held view of the place. The Galápagos Islands were as damned a place as a mid-nineteenth-century Yankee like Herman Melville could conceive. In his ten published sketches about it, he called the archipelago by its old Spanish name, "The Encantadas, or Enchanted Isles," and for Melville it represented "evilly enchanted ground."

3

Perhaps inspired by Dante and Milton, to whom he alluded, Melville peopled his Galápagos with spirits, tormented souls and an "unknown" number of fiends—but none of what he viewed as the higher forms of life.
"Man and wolf alike disown them. Little but reptile life is here found," he
observed. "No voice, no low, no howl is heard; the chief sound of life here
is a hiss." Evil enchanted even these lower beasts. The islands' remarkable
giant tortoises delighted Darwin, but Melville portrayed them as reincar-
nations of wicked sea captains doomed to near-eternal life in a near-
unlivable place.[2]

Melville visited the islands only briefly. Late in 1841, as a twenty-two-
year-old sailor he cruised among them for a month, with occasional an-
chorages, during his apprenticeship aboard the *Acushnet*, the New England
whaler that he called his Harvard College and his Yale. After jumping ship
on a less forbidding island a few months later, he sailed through the Galápagos at least twice more as a seaman on other ships before returning to
Massachusetts in 1844. These voyages became for Melville the source for a
lifetime of stories, novels and poems. The Galápagos Islands appeared in
several of them and were featured in a series of short sketches, first published in *Putnam's Monthly Magazine* in 1854.[3]

Four aspects of the islands doomed them in Melville's mind: harsh terrain, inhuman desolation, striking novelty and changeless permanence.
His vision reflected that of many other early visitors to the islands, including naturalists.

At first sighting, the archipelago still shows the same damned face described by Melville.[4] "Take five-and-twenty heaps of cinders dumped here
and there in an outsize city lot," he began his first sketch, "imagine some
of them magnified into mountains, and the vacant lot the sea; and you will
have a fit idea of the general aspect of the Encantadas." Although largely
in the South Pacific, these islands differed greatly from the lush tropical
ones Melville depicted in his earlier, more romantic tales. Those earthly
gardens of delight and temptation barely touched this cursed place. Here
Melville described "wide level beaches of multitudinous dead shells, with
here and there decayed bits of sugar-cane, bamboos, and cocoanuts,
washed upon this other and darker world from the charming palm isles to
the westward and southward; all the way from Paradise to Tartarus."[5]

Later naturalists would see such beached bits of foreign plants as the source of new species or extensions of old ones—but not Melville. "On most of the isles where vegetation is found at all," he wrote, "it is more ungrateful than the blankness of Atacama. Tangled thickets of wiry bushes, without fruit and without a name, springing up among deep fissures of calcined rock." Much of the shoreline was lava cliffs and stone, lacking even the invitation of a dead-shell beach. "In many places the coast is rock-bound, or more properly, clinker-bound; tumbled masses of blackish or greenish stuff like the dross of an iron-furnace, forming dark clefts and caves here and there, into which a ceaseless sea pours a fury of foam."[6]

The islands varied only in the intensity of their desolation. Melville described volcanic Narborough as "the loftiest land of the cluster; no soil whatever; one seamed clinker from top to bottom." Its central volcanoes, rising some five or six thousand feet and "standing grouped like a gigantic chimney-stack," represented the ninth circle of desolation. "There toil the demons of fire, who at intervals irradiate the nights with a strange spectral illumination." Surrounding Narborough on three sides lay "the black jaws of Albemarle," the largest of the islands. The resulting channel was a rich feeding area for whales and thus attracted great numbers of whalers; but Albemarle remained nearly as desolate as Narborough according to Melville's census, with no human life and "an incomputable host of fiends, ant-eaters, man-haters, and salamanders."[7]

Moving further from this center, Melville found smaller islands where the terrain grew less foreboding. None of these outer islands had a native human population; indeed, "it may be reasonably concluded that they have been immemorial solitudes." Yet he told of scattered runaways and castaways—refugees from passing ships or the South American mainland—who struggled to survive in rock huts on islands with little fresh water or arable land. These were tortured loners, "some of whom too sadly experience the fact that flight from tyranny does not of itself insure a safe asylum, far less a happy home." For Melville, who like most of his contemporaries saw the earth as created for human habitation, such "emphatic uninhabitableness" provided ample basis for doubt "whether any spot of earth can, in desolateness, furnish a parallel to this group."[8]

What life he found contributed to "that air of spell-bound desertness which so significantly invests the isles." Referring to the dearth of mammals, he lamented, "The Encantadas refuses to harbor even the outcast of the beasts." Few stately plants dwelled in "these hot aridities." Strange reptiles, however, abounded on the land: "tortoises, lizards . . . and that strangest anomaly of outlandish nature, the *iguana*." Rather than a natural development on islands devoid of mammalian predators, Melville saw these dominant reptiles as further evidence of the place's curse. "There is something strangely self-condemned in the appearance of these creatures," he wrote of the Galápagos Tortoises, which reportedly can live for more than two centuries. "Lasting sorrow and penal hopelessness are in no animal form so suppliantly expressed as in theirs; while the thought of their wonderful longevity does not fail to enhance the impression."[9]

Even stranger enchantments filled the sea and sky. "Below the waterline, the rock seemed one honey-comb of grottoes, affording labyrinthine lurking places for swarms of fairy fish," Melville wrote of one Galápagos anchorage. "Nothing was more striking than the complete novelty of many individuals of this multitude. Here hues were seen as yet unpainted, and figures which are unengraved." Above his ship rose a towering rock called Rodondo. "And as the eaves of any old barn or abbey are alive with swallows, so were all these rocky ledges with unnumbered sea-fowl," he reported. "All would have been bewitchingly quiescent, were it not for the demoniac din created by the birds." Here Melville became an amateur naturalist: "I know not where one can better study the Natural History of strange sea-fowl than at Rodondo. It is the aviary of Ocean." Penguin, pelican, albatross, gannet, booby and storm-petrel "haunt" this rock in tiers—novelties all, at least to Melville's Yankee eye, though somewhat less so to the science of his day. "With ear splitting cries the wild birds celebrate their matins," he wrote of sunrise at Rodondo. "Each moment, flights push from the tower, and join the aerial choir hovering overhead."[20]

One striking enchantment marked both the birds and sea life here—their complete indifference to the sailors, who could pick them up or knock them down at will. "In your victimized confidence," Melville observed, "you are of the number of those who inconsiderately trust, while they do not understand, human nature."[22] On land, underwater and in air, the animals of the Galápagos neither knew nor were known by humans. This to Melville was sign and symbol of their evil enchantment.

"But the special curse, as one may call it, of the Encantadas," he concluded, "is that to them change never comes." Other places suffer solitude, but the poles had regular seasons and the open sea had set movements that "mitigate their terror"—not so the Galápagos. "Cut by the Equator, they know not autumn and they know not spring," Melville lamented, "while already reduced to the lees of fire, ruin itself can work little more upon them."[22] Even the ocean currents ran with unpredictable force amid the islands, and the region's notoriously light winds could maroon sailing ships for weeks at a time. The Galápagos Islands' harsh terrain, inhuman desolation and strange wildlife became their permanent curse.

In that time before evolutionary thinking enlivened natural history, change did not feature prominently in the scientific worldview: biological species and geological features appeared fixed at best; at worst, they had deteriorated from their created goodness. While this left a seemingly peaceable kingdom in western Europe and North America, softened by human toil and common grace, nothing softened Melville's Galápagos Islands from their fiery creation. "The impression they give to the stranger pulling close up in this boat under their grim cliffs is, that surely he must be their first discoverer," he remembered. "How often . . . did I gaze upon that land, not of cakes but of clinkers, not of streams of sparkling waters, but arrested torrents of tormented lava." Thus he captured the prevailing view of the day: "An archipelago of aridities, without inhabitant, history, or hope of either in all time to come."[23] It still strikes many visitors the same way today, unless an evolutionary perspective enlightens their eyes to see it as a place pregnant with possibilities—an enchantment of evolutionary life.

With the emergence of that more modern view, biologists (if not the casual observer) came to see this "archipelago of aridities" as something akin to hallowed ground. Science, more than any other intellectual force, exorcized the curse that seemed to hold the Galápagos Islands spellbound. No longer could any place in nature hold evil or withstand change.

The slow transformation of worldviews had begun two centuries before Melville's day. Kepler's lawbound solar system, Newton's universal gravity and Descartes's inanimate matter in purposeless motion eliminated the need for supernatural interference in the day-to-day operation of the physical universe—potentially pushing God back to the moment of creation, though Newton retained a clock tender's role for his Divine Clockmaker. Standing on the shoulders of these giants, Enlightenment Age physicists constructed an increasingly mechanical universe, culminating around 1800 in Pierre-Simon Laplace's nebular hypothesis, which posited a purely naturalistic origin for the earth, its solar system and stars in general.

Although Laplace's ideas first came to the United States during Melville's youth, they had little immediate impact on prevailing views of biological origins and development.[24] Kepler, Newton and Descartes had all clung to a belief in indwelling supernatural human souls. Succeeding generations of scientists continued to see vital forces, spirits and souls shaping the creation and form of biological species—though typically not as grotesquely as in Melville's imagination. Still, his characterization of the Galápagos as "apples of Sodom, after touching" roughly reflected the scientific orthodoxy of his day: the islands' barrenness and lack of "higher" species surely displayed their Creator's special disfavor.[25] Yet they displayed quite a different set of enchantments to the naturalist Charles Darwin, on his own youthful voyage of personal discovery aboard the British survey ship HMS *Beagle*.[26]

Darwin visited the islands six years before Melville did, and published a very popular narrative of his travels so quickly that Melville could have read it on his return voyage to the United States. Certainly the shipboard library of Melville's homeward-bound vessel carried the book, which Darwin dryly called his *Journal of Researches*, and Melville later bought his own copy. By the time Melville composed "The Encantadas," Darwin's travel narrative had become something of a classic in its genre, the best known account of the archipelago.[27] Melville quoted from it in his own classic, *Moby-Dick*. Yet even though *Journal of Researches* did not betray its author's still-secret thoughts on naturalistic evolution, it was too spiritually sterile for Melville's tastes. In his Galápagos sketches, he did not list Darwin's book among the "eye-witness authorities worth mentioning touching the Enchanted Isles," but rather lumped it among the "barren, bootless allusions from some few passing voyagers or compilers."[28]

Melville then proceeded to reject several of Darwin's scientific observations in favor of more sentimental ones. Where Darwin, for example, made a point of experimentally refuting the romantic belief that overturned tortoises can never right themselves, Melville took pains to reassert that "everyone knows" just the opposite. Where Darwin reported finding only eleven kinds of waterbirds, including only one type of gull, Melville countered that he found "unnumbered sea-fowl" and "gulls of all varieties."[29]

Melville also parodied Darwin. The *Journal of Researches* featured a chart listing the numbers of different species Darwin saw on various islands of the Galápagos. Melville countered with a mocking chart of his own. "If now you desire the population of Albemarle, I will give you, in round numbers, the statistics," he sneered. He then listed the numbers as "unknown" for most types and gave gross figures for the rest, "making a clean total of 11,000,000."[20] For Melville, one literary critic concluded, "the scientific method of quantitative analysis for purposes of investigating the nature of the world is absurd since the ultimate nature of the world is finally 'unknown.'"[22]

Melville's and Darwin's contrasting accounts of Galápagos Tortoises marked an epochal change in worldview. *Journal of Researches* provided an objective description of these locally dominant reptiles, setting forth their habits, diet, size and distinguishing characteristics. Melville focused on their subjective meaning and purpose: the "spectre-tortoise," he called them, enthroned in the archipelago as the "sole solitary Lords of Asphaltum." Whereas Darwin dispassionately noted the tortoises' instinct of fixedly moving toward their goals, Melville characterized this trait as a "crowning curse" that caused them to butt futilely against immovable obstructions in a fashion reminiscent of the punishment reserved for hoarders and wasters in the fourth circle of Dante's *Inferno*. "That these tortoises are the victims of a penal, or malignant, or perhaps a downright diabolical enchanter, seems in nothing more likely than in that strange infatuation of hopeless toil," he observed, "their drudging impulse to straightforwardness in a belittered world." The tortoise's longevity existed within natural time spans for Darwin, while for Melville it extended to "dateless, indefinite endurance."[22]

Name of Island.	Total No. of Species.	No. of Species found in other parts of the world.	No. of Species confined to the Galapagos Archipelago.	No. confined to the one Island.	No. of Species confined to the Galapagos Archipelago, but found on more than the one Island.
James Island .	71	33	38	30	8
Albemarle Island	46	18	26	22	4
Chatham Island	32	16	16	12	4
Charles Island .	68	39 (or 29, if the probably imported plants be subtracted)	29	21	8

Darwin's chart of Galápagos species, from Darwin, Journal of Researches, *2d ed. (1845)*
(Source: Used by permission of UGA.)

Melville's tortoises embodied the changelessness of the pre-Darwinian world. "Ye oldest inhabitants of this, or any isles," he wrote, "what other bodily being possesses such a citadel wherein to resist the assaults of Time?" The Galápagos Tortoise thus appeared not merely cursed but eternally cursed, much like the islands themselves. "In no world but a fallen one could such lands exist," Melville concluded—a concept utterly alien to the emerging naturalism of modern science.[22]

Evolutionary naturalism, devoid of souls and curses, took shape within the scientific culture even as Melville wrote within the literary one. In a telling coincidence, the first of Melville's Galápagos sketches appeared in the same issue of *Putnam's* as a review of a new book by the British naturalist Alfred Russel Wallace, who soon gained fame as codiscoverer (with Darwin) of the theory of organic evolution by natural selection. Although Wallace's book dealt solely with his South American field research, it contained hints of the new theory, and the review incorporated reports of intermediate types linking species.[24] Another article in the same issue— *Putnam's* billed itself as a "magazine of literature, science, and art"—discussed new concepts of geologic evolution revealing an earth "older and

If now you desire the population of Albemarle, I will give you, in round numbers, the statistics, according to the most reliable estimates made upon the spot:

Men, ...	none.
Ant-eaters, ...,	unknown.
Man-haters, ... , .	unknown.
Lizards, ...	500,000.
Snakes, ...	500,000.
Spiders, ...	10,000,000.
Salamanders, ...	unknown.
Devils, ... : . . .	do.
Making a clean total of	11,000,000.

exclusive of an incomputable host of fiends, ant-eaters, man-haters, and salamanders.

Melville's Chart of Galápagos populations from Putnam's Monthly Magazine *(1854) (Source: Used by permission of UGA.)*

grander than the uninstructed mind of man ever imagined," and proposed using the fossil record "to trace back the long order and sequence of the past." The requisite ancient fossils should exist, the essayist suggested, "In some quarter of the globe where volcanic fires burn fiercest, where their forces have depressed the land beneath the sea, and lifted up the ocean-bed to become dry land,—perhaps on the coast of Chili or among the islands of the Pacific." The secrets of life were thus to be sought on islands much like those described by Melville sixty pages later as bearing death's curse.[25]

Within a decade, emerging concepts of biological and geological development merged into a comprehensive theory of organic evolution by natural selection. South American biology and Pacific Rim geology would play key parts in Darwin's case for a purely naturalistic origin of organic species. Such thinking directly challenged traditional notions of purpose and meaning in life, and escalated the culture wars that pitted science against religion, and reason against revelation.[26]

As the impact of Darwin's vision grew within Western science and culture, so did the role of the Galápagos in breaking the spell of enchantment on the origin of species. The islands became a field laboratory for the study of evolution in action, their harsh environment an opportunity rather than a curse. Even the tortoises metamorphosed from damned souls into an endangered species.

Ever since those days when Darwin walked on the Galápagos Islands with his eyes wide open in wonder, the archipelago has played an ever larger role in the history of science. "No area on Earth of comparable size has inspired more fundamental changes in Man's perspective of himself and his environment than the Galápagos Islands," notes Robert Bowman, the current dean of Galápagos researchers. "The unconventional flow of nature on the land and the sea has spawned revolutionary views on the origin, not only of new species, but also of life itself."[27]

Eco-tourists followed scientists in seeking to see what Darwin saw, threatening those very sites by their numbers and transforming the Galápagos into a test case for environment protection from the people who love it too much. The archipelago became Ecuador's first national park, a founding "Man in the Biosphere" preserve for UNESCO and one the United Nations' initial World Heritage Sites. Tens of thousands of Ecuadorian settlers followed hundreds of thousands of tourists to this once isolated archipelago, compounding the human pressure on its fragile ecosystem and leading to such previously unthinkable episodes as the accidental grounding in 2001 of a fully loaded oil tanker almost within sight of the intersection of Charles Darwin Avenue and Herman Melvile Street in the Galápagos's bustling capital.

A half century ago, Julian Huxley spoke of the islands' past and present significance. "The Galápagos Archipelago is historically of great scientific importance, since it was its fauna and flora which more than anything else convinced Charles Darwin of the fact of evolution," Huxley wrote of the archipelago's past. Of the present he then added, "It provides indeed one of Nature's most clear-cut experiments in evolution, and for this reason, and as a memorial to Darwin's great achievement, its flora and fauna should be studied, preserved and safeguarded."[18] The history of the Galápagos in science and human thought from the past to the present is the story of this book; increasingly, the Archipelago's future is the story of us all.

PART ONE

CREATIONIST CONCEPTIONS

ONE

CURSED BY GOD
AND NATURE

REMOTE ISLANDS CAN PUZZLE SCIENTISTS, and none have puzzled them more than the Galápagos. How, for example, did the native plants and animals get there? Consider the alternatives. Suppose God created all plant and animal species at one time and in one place, as suggested by the Book of Genesis. How then were remote islands stocked with plants and animals—particularly those types that seem unable to swim, fly or float from the mainland? Or suppose there were multiple creations in different places or, more simply, that species always existed where they now reside. Why then are island species similar to, but not necessarily the same as, mainland ones? Finally, suppose current species developed from preexisting organic forms by some natural or supernatural mechanisms. What then do the island species reveal of the larger scheme?

Scientists have considered all these alternatives. Among the ancient Greeks, for example, Aristotle assumed the eternity of species while Anaximander suggested evolutionary origins. Yet for Western science the issue remained largely academic until the sixteenth century, when voyages of discovery began broadening European horizons to include new worlds and remote islands. Then the question of origins became inescapable for European scientists seeking to understand the strange species they found. Nowhere did they find stranger species than on the Galápagos.

The archipelago actually posed a puzzle within a puzzle. The larger puzzle was America.

Convinced that he had reached the Far East by sailing west, Christopher Columbus never fully comprehended what he found. "No animals of any kind," he wrote of his first American landfall in 1492—clearly meaning no land mammals. Four days later, after a more complete reconnaissance, he repeated in his journal, "I have seen no land animals of any kind except parrots and lizards." He interpreted this to mean simply that he had not yet reached the Asian mainland, where he knew many land mammals lived. Yet he wondered about his other discoveries. "Fish of such different kinds from ours that it is a marvel," he observed, and "trees of very different kinds from ours." Fish could swim to these islands and seeds could float—so the presence of sea life and plants made sense to him—but why so unlike European kinds? He soon made the observation universal: "The trees are as different from ours as the day from the night, and the fruits too and the grass and the stones and everything else." He could only attribute their unfamiliarity to his ignorance of Asia. "I am the saddest man in the world at not recognizing them," he lamented in his journal, "because I am certain that they are all of value."[1]

As he extended his explorations in the "Indies," Columbus began finding some familiar types, but nagging differences remained. On Cuba, for example, his sailors "saw no four-legged animals except dogs that did not bark." As he approached Hispaniola, which became his base for operations, Columbus happily noted that "a mullet just like those in Spain jumped into the boat"—the first local fish that he recognized. Only on a later voyage, when he reached the South American continent, did he finally encounter large mammals—evidence, he thought, of his having at last found the Asian mainland; but even these animals, which he compared to European deer, boar and wildcats, struck him as "very different from ours."[2]

Columbus's mute Caribbean dogs symbolized the challenge that American nature posed for European naturalists. Since medieval times, Europeans had seen the natural world as a vast spiritual allegory created by God to instruct humans—a kind of tangible revelation. Perhaps the best known of these allegories involved dogs, of which St. Ambrose wrote, "To dogs is given the ability to bark in defense of their masters and their homes. Thus you should learn to use your voice for the sake of Christ, when ravening wolves attack His sheepfold." Under this view, muteness deprived dogs of both function and instructive value, making their presence all but incomprehensible. Other American animals raised similar difficulties. "Prior to the age of discovery, a symbolic world had existed in which a discrete set of natural objects had provided a ground for the composition of unlimited variations on eternal themes," historian Peter Harrison notes. "Now, however, what had once been a coherent universal language was inundated by an influx of new and potentially unintelligible symbols." The very structure by which Europeans understood nature began to collapse under the collective weight of exotic animals and plants.[3]

Collapse took time, however. Following Columbus, other Spanish explorers returned to Europe with similar reports. Plants, birds, reptiles and fish abounded, but inevitably they differed from Europeans kinds; and still there remained a disquieting scarcity of large land animals. From Central America, conquistador Martín Fernàndez de Enciso wrote of "tailed cats, which are like monkeys except that they have long tails," and of alligators as big as "bullocks." Of the latter he proudly added, "I myself succeeded in killing the first that was killed." From North America, Giovanni da Verrazzano told of "wild animals that are considerably more wild than in our land of Europe." Yet neither explorer reported many large mammals; mostly rabbits, deer, wolves and the like—none of which impressed them. "This bitter observation," historian Antonello Gerbi wrote, "was to hang like a millstone on the fauna of all America over the ensuing centuries. It seemed to point to some organic inferiority or deep-rooted poverty, and in some cases prompted the formulation of theories of physical decadence or immaturity." In Rome, the Pontiff characterized the native American peoples as vegetarian on the assumption that they had no proper meat to eat. Inferiority and decadence would dominate European views of the Americas until the 1800s.[4]

Among the first to recognize Columbus's Indies as part of a New World, Amerigo Vespucci firmly asserted that the territory contained plants and animals unknown to Europe. Together with Old World types, they would constitute so many different kinds that he wondered how a pair of each could have fit on Noah's ark. Indeed, he added, "If I have to attempt to write all the species of animals [in America], it would be a tedious tale." Tedious perhaps, but not repetitive of European listings. Referring to the ancient Roman encyclopedist Pliny the Elder, who assembled over 20,000 separate observations about nature into a massive "natural history" of the world that (in 1500) still seemed comprehensive to many in Europe, Vespucci asserted about America, "I believe certainly that our Pliny did not touch upon a thousandth part of the animals and birds that exist in this region."[5]

Once Europeans realized that America represented a vast new continental land mass, some sought to become its Pliny. Adopting the role of a medieval herbalist, Nicolas Monades labored in Spain to assemble reports of New World plants supposed to have medicinal properties. Supplementing collected reports with personal observations, Gonzalo Fernández de Oviedo and José de Acosta composed ambitious natural histories of the New World. Like Pliny, these sixteenth-century Spaniards sought more to provide useful and entertaining information about nature than to offer reasoned explanations of its operation.

Following medieval tradition, Oviedo's chief interests in living things were utility, nuisance value and novelty. Thus he opened a chapter on "beasts" in his *Natural History* by announcing, "to let pass the multitude of things which are as variable as the power of nature is infinite, and to speak of such things as come chiefly to my remembrance as most worthy to be noted, I will first speak of certain little and troublesome beasts which may seem to be engendered of nature to molest and vex men." He began with leeches and passed on to fleas, snakes, spiders, toads, crabs and iguanas. Typically these "seemed very strange and marvelous to the Christian man to behold, and much different from all other beasts which have been seen in other parts of the world." Oviedo described spiders "bigger than a man's hand," American toads larger than European house cats and iguanas "fearsome to sight." Far from disparaging the American fauna, he reveled in its strangeness; but he found few stately New World mammals, and this clearly concerned him. He denounced the American lion as "cowardly" and tried in vain to urge a proper bark out of Caribbean dogs.[6]

Acosta followed Oviedo in taking a Christian creationist's anthropocentric view of nature in the New World. "In our discourse of plants we will begin with those which are proper and peculiar to the Indies," he wrote to open a typical chapter of his *Natural History*, "and forasmuch as plants were chiefly created for the nourishment of man, and that the chief (whereof he takes his nourishment) is bread, it shall be good to show what bread the Indians use." In a later chapter on "notable fowls," Acosta added that in Mexico "there are abundance of birds with excellent feathers, so as there be not found in Europe that come near them." Having "been so appointed by the sovereign Creator for the service of man," he added, these notable American birds provided to humanity "not only their flesh to serve for meat, their singing for recreation, their feathers for ornament and beauty, but also their dung serves to fatten the ground."[7]

Acosta wondered about how all these useful beings came to the New World without leaving any of their kind in the Old World. He presented two problematic options. God could have created them in America after the Flood, Acosta offered, but then "we could not affirm that the creation of the world was made and finished in six days." While this was clearly heresy, the orthodox alternative posed problems as well. "If we say then that all these kinds of creatures were preserved in the Ark by Noah," Acosta asked, "I demand how it is possible that none of their kind should remain here [in the Old World]?" Forced by his Christian faith to accept the second option, he concluded, "We must then say, that though all beasts came out of the ark, yet by a natural instinct and the providence of heaven, diverse kinds dispersed themselves into diverse regions, where they found themselves so well, as they would not part, or if they departed, they did not preserve themselves, but in the process of time, perished wholly." This became the Church-approved answer, but it stretched credibility even then. Scientific-sounding freethinkers such as Giordano Bruno enjoyed picking at this particular sore—at least until they were silenced, as Bruno was, by the Inquisition.

Oviedo, Acosta and the other early Spanish chroniclers retained a distinctly medieval approach toward nature. Their botanical writings follow the herbalist tradition of identifying plants and describing their human uses. Their zoological observations read like entries in medieval bestiaries, which typically included instructive and entertaining accounts about various animal types, chiefly derived from Pliny and following the classifications of Aristotle. The Copernican revolution did not begin transforming European concepts of celestial physics until fifty years after Columbus first encountered America, and traditional concepts of terrestrial biology were affected even later. "The so-called scientific revolution of the sixteenth and seventeenth centuries," the great zoologist Ernst Mayr noted, "caused no change at all in the attitude toward [biological] creationism."[8] In the meantime, religiously influenced concepts of a created world designed to nurture and instruct humans (and to vex them as needed) shaped the European understanding of the New World.

In 1529, when Francisco Pizarro's looting of Inca gold began drawing Europeans south toward Peru and the Galápagos, they brought their medieval conceptions of nature with them. "The discovery of South America burnt upon Europe like a fabulous dream," early twentieth-century Galápagos travel writer Victor von Hagen noted. Yet like a dream, the new continent proved difficult to reconcile with supposed reality. In his 1590 *Natural History*, Acosta obviously labored to assign berths for Peruvian animals on Noah's ark and to find divine purposes for them in America.[9]

Following the opening of silver mines in Peru, an apparent desire to quash theological questions raised by New World discoveries, coupled with a real concern to exclude other Europeans from its rich domains, led Spain to restrict scientific study in America and suppress reports about it. The Spanish government seemed indifferent to most aspects of its American possessions beyond their mineral wealth and native workers, and rarely even requested detailed reports from its America-bound explorers and sea captains. After other European nations began sending explorers into the Caribbean basin and along the Atlantic coast of the Americas, Spain clamped down still tighter on its Pacific Coast dominions and severely restricted travel to its richest possessions. Peru became El Dorado, the stuff of fabulous legend.[10]

If its novel species made South America difficult to fit into traditional European conceptions about nature, then the Galápagos Islands compounded those difficulties immensely because of a missing one. Nearly every island of any size or significance already had people living there when Europeans discovered it. "Discovery" was simply a Western convention to legitimize the conquest of these places in the eyes of other Europeans. The Galápagos stand as one of the rare exceptions to this rule. Despite vague Inca legends to the contrary, no one has ever detected any credible evidence that humans lived on these islands before the first European ship entered their waters, in 1535.[11] Their virgin state would one day help make the Galápagos Islands especially intriguing to scientists. At the

Drawing of a Galápagos island, from Edward Cooke, A Voyage to the South Sea *(1712)*
(Source: Used by permission of UGA.)

time, such a state rendered them of little interest to the Spanish. If God made the earth for human habitation, they wondered, what could explain an uninhabited place? Fittingly, a Roman Catholic bishop discovered these islands (one of the few times that a church official played such a role) and promptly pronounced them cursed of God. Thereafter, pious Spaniards generally avoided them.

After they had brutally subjugated the native people of Peru by 1532, Pizarro and his soldiers turned on each other. The Spanish government then directed the bishop of Panama, Fray Tomàs de Berlanga, whose diocese included Peru, to investigate both the bloody suppression of native Peruvians and the ongoing power struggle among the conquistadors. Outfitting a vessel, Berlanga sailed with men and horses along the Pacific coast from Panama toward Peru. "Something," as he put it, then drew him to the unknown archipelago offshore.[12]

"The ship sailed with very good breezes for seven days," Berlanga later reported to the king, but then encountered "a six day calm." With no wind, a strong westward current "engulfed us in such a way that on Wednesday, the tenth of March, we sighted an island." Soon they spotted several more, some quite large. Unable to measure longitude at sea or gauge distances traveled when carried by a current, Berlanga had no idea just how far off course his ship had drifted. He wrote "that we were not more than twenty or thirty leagues from the soil of Peru," when in fact five hundred miles separate the Galápagos Islands from the South American mainland. He could calculate the islands' latitude, however, and found that "they are between half a degree and a degree and a half south latitude." Berlanga had discovered the archipelago's southern end, just below the equator.[13]

The islands offered poor haven. When some sailors landed on the first island to look for much needed fresh water, according to Berlanga, "they found nothing but seals, and turtles, and such big tortoises, that each could carry a man on top of itself, and many iguanas that are like serpents." Growing desperate for drinking water, the entire crew disembarked on the second island, "thinking that on account of its size and monstrous shape, there could not fail to be rivers and fruits." Again they suffered disappointment. "Some were given the charge of making a well," Berlanga wrote, but "from the well there came water saltier than that of the sea." Others looked for freshwater streams or ponds, he added, but "they were not able to find even a drop of water for two days and with the thirst the people felt they resorted to a leaf of some thistles like prickly pear, and because they were somewhat juicy, although not very tasty, we began to eat of them." Water squeezed from these cactus pads "looked like slops or lye," but the party "drank of it as if it were rose water." Their penance on the islands apparently completed, the bishop then reported that after he said mass on Passion Sunday, "the Lord deigned that they should find in a ravine among the rocks as much as a hogshead of water, and after they had drawn that, they found more and more." Two sailors and ten horses had died of dehydration, but God saved the rest. They managed to sail back to the mainland, and never returned.[14]

Berlanga's report damned the archipelago. "I do not think that there is a place where one might sow a bushel of corn, because most of it is full of very big stones," he wrote, "and the earth that there is, is like dross, worthless, because it has not the power of raising a little grass, but only some thistles." The fauna was as wretched as the flora: "many seals, turtles, iguanas, tortoises, many birds like those of Spain, but so silly they do not know how to flee, and many were caught in the hand." God did not make this place for humans, the bishop concluded. "It seems as though some time God had showered stones."[15] Following this report, the Spanish government never attempted to colonize the islands and did not even investigate them again for nearly two centuries.

The contrast between Berlanga's Galápagos report and Columbus's initial New World journal entry is telling. Although expressing his disappointment in not immediately finding valuable minerals, Columbus boasted of "well built" native people "with strong bodies and fine features" who "ought to make good slaves" and "could very easily become Christians."[16] In Peru, Pizarro found both minerals and native peoples to exploit—making it the most valued of all Spain's overseas possessions. Scientific interests followed economic ones: in his *Natural History*, for example, Acosta focused on the mineral wealth and people of Peru—Spain's treasures. Spanish America's greatest naturalist, Alessandro Malaspina, would do the same a century later. Yet the Galápagos Islands offered neither mineral wealth nor exploitable natives. In the most telling omission of his report to the Spanish king, Berlanga did not even bother to name them. Unlike the created beings given names by Adam in the Genesis account, these islands were not good.[17]

A name, when it was finally given, was inspired by the terrors of a second accidental visit—also resulting from continuing conflict on the mainland. Berlanga's mission failed to end the warfare among the conquistadors of Peru, and after berating Pizarro for mistreating the native people, the bishop resigned his position and returned to Spain. A decade later, in 1546, a Spanish officer named Diego de Rivadeneira fled by boat toward Central America with a dozen solders and a few impressed sailors from the vengeful forces of Pizarro's brother, Gonzalo. Spanish ships heading north from Peru normally hugged the coast, passing far to the east of the Galápagos—but not Rivadeneira's leaky vessel. Keeping well outside South American coastal waters to avoid Pizarro's ships, he sighted islands that he recognized as Berlanga's archipelago.

Rivadeneira spotted one large island first, which he described as "covered by a cloud" with "high mountains near the coast." Some of his crew recalled seeing smoke rising from the mountains, the first written record of volcanos on the Galápagos Islands. Low on drinking water, Rivadeneira attempted to land his boat on the large island. The strong, unpredictable Galápagos currents took hold of his craft, however, and for three days kept it shifting about within sight but beyond reach of land. Other islands came into view during these days but also proved elusive. From Rivadeneira's shipboard vantage point, the islands themselves seemed to move about so much that he called them *Las Encantadas*, the Enchanted Isles, and referred to their "apparent fleetingness and unreality."[18]

The islands proved no more welcoming when Rivadeneira finally managed to land on one of them. After the thirsty soldiers and sailors disembarked, the official chronicler reported, "they set out in different directions to look for some water. But each fearing that he would be left behind by the others, they soon came back to the shore and re-embarked to continue on their way very sadly because of their lack of food and water."[19] Not appreciating the great distance from the archipelago to the mainland, Rivadeneira then set sail without water on a desperate race to Central America.

Following his arrival in Guatemala after a harrowing voyage, the official expedition report to the Spanish government, like Berlanga's ten years earlier, focused on the archipelago's distinctive animals. According to this brief account, Rivadeneira "found tortoises, turtles, iguana, sea lions, birds called flamingos, doves and other birds, among them a beautiful gyrfalcon which has never been seen here or I believe in Peru."[20] Unusual animals meant little to Spanish officials. The government ignored Rivadeneira's request to explore and perhaps settle the islands, and sent no other explorers or settlers there for more than 150 years.

Although these chance encounters kindled little interest among the Spanish, they brought the archipelago to the attention of Europeans generally. Seventeenth-century European cartographers, who competed to publish maps showing the latest "discoveries," eagerly scoured accounts from explorers and travelers returning from the New World. The great Flemish cartographer Abraham Ortelius first entered the islands onto maps published in Europe, based on an agent's reading in Spain of official reports from Berlanga's voyage. Because the great tortoises (*los Galápagos*, in Spanish) stood out as the islands' distinguishing characteristic in these reports, Ortelius identified the archipelago in his classic 1570 *Orbis Terrarium* as "Insulae de los Galopegos," which became "Isolas de Galapagas" on his maps by 1574.[21] That 1574 map showed only one major island at the site given by Berlanga and two adjacent islets, far from the "ten or twelve" islands noted by Rivadeneira.[22] It sufficed, however, to locate the archipelago in the European mind, and that location suggested its potential as a base for raiding Spanish ships and ports along the rich Peruvian coast. Buccaneers opened the next phase of European contact with the Galápagos.

At the time, buccaneers held a shifting status somewhere between adventurers and thieves. Their occupation developed in the wake of Spain's conquest of Mexico and Peru. Beginning early in the 1500s, vast amounts of gold and silver poured from those two places (particularly the latter) through the Caribbean to Spain. Simultaneously, England and France emerged as rival maritime powers intent on profiting from the commerce in precious metals from the New World. The Netherlands soon joined them when it rebelled from under Spanish rule. Ships from these rival nations, often captained by adventuresome courtiers or ambitious naval officers, regularly invaded Spanish American waters during the sixteenth and seventeenth centuries. The invaders typically sought contraband trade with Spanish colonists or plunder on land and sea, whichever might prove easier. Occasionally they attempted to establish or conquer island bases or mainland colonies, such as Jamaica and Guiana. During periods of open war between their home country and Spain, the raiders often served legally as privateers with official government commissions. At other times they acted independently as pirates, more or less condoned by their own governments. "The result could be much the same to a Spanish colonial if he were taken by a man of war, a privateer, an armed smuggler, or a pirate," one historian notes. "The goal of all was Spanish treasure and the humbling of Spanish pride."[23] In time, they all became known as buccaneers.

Most buccaneer activity occurred in the Caribbean, but some spilled over into the Pacific. The Caribbean offered easy access from Europe and refuge on English, French and Dutch-held islands. The competition among buccaneers for treasure passing through Caribbean waters became so fierce, however, that some sought it nearer its source, knowing that it passed from Peru up the Pacific coast before crossing the isthmus of Panama and entering the Caribbean. The English sea captains John Oxenham, Francis Drake, Thomas Cavendish and Richard Hawkins showed the way into the Pacific during the final quarter of the sixteenth century. Drake, at least, returned to England in glory, his ship loaded with Peruvian treasure, after circumnavigating the globe. Although each of these early English adventurers made use of islands off the Pacific coast of South

America, none of them sought refuge as far offshore as the Galápagos—though it might have saved Oxenham and Hawkins from capture. In his journal, Hawkins dismissed the archipelago with a one-line entry as he sailed abreast of it: "Some fourscore leagues to the westward of this cape lyeth a heape of Illands the Spaniards call Illas de los Galápagos; they are desert and bear no fruite."[24]

After a half century in which Dutch sailors dominated privateering in the region during their country's long war of independence from Spain, a second wave of English buccaneers entered the Pacific during the 1680s. The Galápagos became a favored refuge for these adventurers, including several who later wrote popular books about their exploits. Four of these literary buccaneers sailed aboard a "lovely" Danish merchant ship that they illegally seized on the high seas off West Africa and renamed *Batchelor's Delight* for service in Pacific piracy from 1684 to 1688.[25] From a naturalist's perspective, William Dampier wrote the best of these books: it earned him entree to England's premier scientific association, the Royal Society of London.[26] Lionel Wafer's book featured detailed observations of useful plants, animals and people from Central America (where he recuperated for months from an injury), but less about the seemingly useless Galápagos. William Cowley and Basel Ringrose added descriptive narrative and swashbuckling adventure. Taken together, the buccaneer journals introduced the Galápagos to Europeans in terms that stirred little immediate interest. The islands seemed destitute heaps of infernal creation. As popular and scientific interests changed, however, these same descriptions helped draw scientists and adventurers to the archipelago as a pristine place of wondrous novelty.

Dampier's *A New Voyage Round the World*, first published in 1697, when circumnavigation remained a notable feat, proved the most enduring and intriguing of the lot. Dampier offered precise observations of remote Pacific Ocean regions, within a tale of murderous piracy on the high seas. The book immediately became widely popular, running through four English and three foreign editions in its first few years.[27] In it, the Galápagos Islands first appear as a hideout after Dampier's band has captured three merchant vessels off the Peruvian coast. "It was the 31st day of May [1684] when we first had sight of the Islands *Gallapagos*," he wrote. "They are of a good height, most of them flat and even on the top; 4 or 5 of the Eastermost are rocky, barren, and hilly, producing neither Tree, Herb, nor Grass, but a few Dildo-trees, except by the Sea side." The islands grew scarcely more inviting on closer examination. Of the so-called dildo-tree (or candelabra cactus), for example, Dampier soon added, "This shrub is fit for no use, not so much as to burn."[28] After an initial two-week stay and occasional subsequent visits, which provided ample opportunity for the energetic Dampier to survey the place, he all but concluded the same about the archipelago as a whole: fit for no use, not so much as to burn.

In the context of Dampier's time and purpose, this was a harsh judgment indeed. Dampier dedicated his book to Charles Mountague, the president of the Royal Society, and eagerly sought to conform it to that organization's scientific mission. In its 1663 charter, the Royal Society took as its purpose "the advancement of the knowledge of natural things and useful arts by experiments, to the glory of God the creator and for application to the good of mankind."[29] Both elements of this purpose—finding in nature both divine design and human utility—reflected the influence of England's lord chancellor and leading philosopher of science, Francis Bacon, whose radical Protestantism inspired him to view the English Reformation as an opportunity for restoring humans to an Edenic position of total knowledge of and dominion over nature. "The true end of knowledge," Bacon once wrote, "is a restitution and reinvesting (in great part) of man to the sovereignty and power (for whensoever he shall be able to call the creatures by their true names he shall again command them) which he had in his first state of creation."[30]

Thus inspired, Royal Society naturalists sought to remake science. God still designed nature, they maintained, but instead of conveying allegorical spiritual instruction through creation, His purpose was to serve human material wants and needs. Science thus imposed order on the world not by passively seeking supernatural meanings but by actively finding practical uses. As part of their program, these Baconian scientists assumed the task of compiling new natural histories of their surroundings. They believed that prior naturalists, distracted by a misguided search for allegorical meaning in the created world, had not looked at nature with an open mind. The new scientists sought to remedy this through comprehensive surveys of natural phenomena aimed at assembling useful knowledge about God's creation. "The study of the natural world thus [remained] a religious activity, albeit in a new sense," Harrison writes. "God's purposes in the creation could only be realised when the functions of those things originally designed for human use were discovered." And damn any things without such use.[31]

Dampier's remarkably detailed accounts of his buccaneering voyages show that he fully subscribed to this program.[32] Professing "a hearty Zeal for the promoting of useful knowledge, and of any thing that may never so remotely tend to my Countries advantage," he wrote in the dedication of *A New Voyage Round the World*, "I must own an Ambition of transmitting to the Publick . . . these essays I have made toward those great ends."[33] Dampier's best biographer calls him a "*natural* scientist," an apt designation in at least two senses: Dampier compensated for his lack of formal training in science with a native knack for keen observation.[34]

Dampier's *New Voyage* exemplified Baconian science: it featured detailed descriptions of the structure, habits and uses of various "sorts" of plants and animals loosely connected in a travelogue format. His account of the coconut, which he hailed as the fruit "most generally serviceable to the conveniencies as well as the necessities of humane Life," ran over four pages. The plantain, which he called "the King of all Fruit, not except the Coco itself," merited similarly detailed treatment, as did breadfruit, cocoa, pineapple, "divers sorts" of citrus and spices, rice and scores of other useful plants. None of these, however, grew on the Galápagos. Dampier dismissed in a single anthropocentric sentence the archipelago's useless dildo-tree: "It is as big as a man's Leg from the root to the top, and it is full of sharp prickles, growing in thick rows from top to bottom"—clearly he had felt its sting. Except for the lush offshore sea grasses that attracted great sea turtles to the area, Dampier dismissed the Galápagos flora as "unknown to us" and apparently useless. The presentation contrasts sharply with his extended analysis of plants found at his other landfalls in the Pacific, including such seeming tropical paradises as Guam, the Philippines, Java and the Spice Islands.[35]

Dampier devoted greater attention to the Galápagos's distinctive animals, but utility remained uppermost in his mind. Having discussed them in detail ten pages earlier in the course of describing his landing at the Juan Fernandez Islands off the Chilean coast, Dampier simply noted the presence on the Galápagos of fur-bearing seals ("a blow to the Nose soon kills them") and enormous sea lions ("all are extraordinarily fat: one of

Galápagos Sea Turtle, from Edward Cooke,
A Voyage to the South Sea *(1712)*
(Source: Used by permission of UGA.)

them being cut up and boil'd, will yield a Hogshead of Oil, which is very sweet").[36] He focused instead on iguanas, tortoises and sea turtles, particularly the latter two.

"The Spaniards when they first discovered these Islands, found multitudes of Guanoes and Land-turtles or Tortoise, and named them the Gallapago's Islands," Dampier noted. "The Guanoes here, are as fat and large, as any that I ever saw; they are so tame, that a man may knock down 20 in an hours time with a club. The Land-turtle are here so numerous, that 5 or 600 men might subsist on them alone for several months." The buccaneers did just that on several occasions and typically, before each departure from the Galápagos, loaded their ships with scores of live tortoises to serve as fresh meat on their voyages. "No Pullet eat more pleasantly," Dampier happily observed. Tortoises lived elsewhere too, of course, and he duly reported finding them in many places, "but whether so big, fat, and sweet as there, I know not." On the Galápagos, they grew up to 200 pounds in weight and "2 feet 6 inches over the Gallapee or Belly," he reported. "I did never see any but at this place, that will weigh above 30 pound weight."[37]

The enormous sea turtles that feed in Galápagos waters so attracted Dampier's attention that he chose this point in this narrative to insert a detailed comparison of the various kinds of turtles encountered on his voyage: the truck-turtle, the loggerhead, the hawksbill and the green. As in his account of tortoises, he dwelt on the striking differences separating the same general types of animals living in different regions. "The Turtle of these Islands Gallapagos, are a sort of bastard green Turtle; for their shell is thicker than other green Turtle in the West or East Indies, and their flesh is not so sweet," he observed. "They are larger than any other green Turtle; for it is common for these to be 3 or 4 foot deep, and their Gallapees or bellies 5 foot wide: but there are other green Turtle in the South Seas that are not so big as the smallest Hawksbill."[38] Certainly his voyage around the world had sensitized Dampier to subtle variations among similar animal types, just as such a journey would be for Charles Darwin 150 years later, but for now each type remained a separate creation ideally designed for its locale and providently set on earth to serve humanity.

In fitting Baconian fashion, Dampier assessed the archipelago's suitability for human habitation and found it wanting. The terrain, "'Tis Rocky and Barren," he noted of the islands he visited, and "here is no [fresh] water but in Ponds and holes in the Rocks." Even getting ashore could prove difficult. Coastal cliffs offered few sites for landing and the sea floor typically fell away "so sharp that if an anchor starts it will never hold." To their credit, however, the islands provided excellent fishing, and Dampier found "the heat not so violent here as in most places near the Equator." Fierce tropical rains pelted the Galápagos from November through January, he reported, but only "moderate refreshing Showers" in spring and fall while "in May, June, July, and August the weather is always very fair." Except for omitting any description of local birds beyond a brief mention of their peculiar tameness, Dampier provided a fair description of the Galápagos for his time, but one unlikely to attract many settlers.[39] Although Dampier's narratives helped to open British eyes to the promise of Pacific exploration and colonization in general, his account of the Galápagos suggested few such possibilities.[40]

Other buccaneer journals filled out the story of privateering in the Galápagos but added little about its natural history. Two Dampier associates, Cowley and Wafer, discussed the buccaneers' 1684 layover in published accounts. Both mentioned the plentiful tortoises, fearless birds and lack of fresh water. Describing the effects of volcanic activity on the terrain without recognizing it as such, Cowley added "that most of these islands having had sulphurous matter that hath sett them on fire, they have been burned formerly and some parts of them blowed up," leaving a desolate landscape strewn with brimstone cinders. He dismissed those he visited as "barren dry islands."[41]

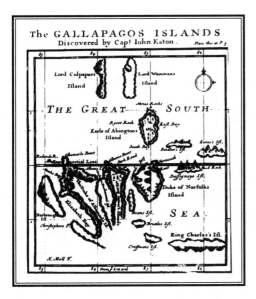

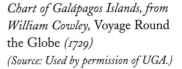

Chart of Galápagos Islands, from William Cowley, Voyage Round the Globe *(1729)*
(Source: Used by permission of UGA.)

Rather than explore this forbidding landscape with Dampier, Cowley spent the buccaneers' layover charting and naming the islands from aboard the *Batchelor's Delight*. English cartographer William Hacke later used Cowley's charts to compose the first general map of the archipelago. Following Cowley, it named individual islands for British nobles and officials friendly to the buccaneers, including major ones for Charles II, James II, the Duke of Albemarle and Sir John Narborough, that remained in common use long after the Ecuadoran government reassigned those islands Spanish names. One tiny speck on Hacke's map bore the playful title "Cowley's Enchanted Island," reminiscent of a passage in Cowley's journal recounting the response when the buccaneers first told their Spanish prisoners of the plan to seek haven on the Galápagos Islands. "The Spanish laugh[ed] telling us that they were inchanted islands," Cowley recalled, "and that they were shadows and not real islands."[42] Cowley's navigation charts finally made them real, and eventually opened them to commerce.

Despite the buccaneers' bleak descriptions, some later sailors followed them to the Galápagos, at least in part because Dampier and Cowley gave credence to secondhand reports of fresh water there. Cowley wrote on faith about "extraordinary good islands where a ship in distress may have anything that can be expected from places not inhabited for there is both food and water and firewood and timber."[43] Dampier and company, visiting the islands after a period of exceptionally heavy rains, had ultimately found sufficient fresh water for their needs from one intermittent spring on James Island. Sailors who followed them rarely had such luck.

After a brief interlude of European peace, English buccaneers returned to the waters off Peru in 1709. Dampier again served as navigator for a company of privateers and twice guided them back to refuge on the Galápagos following successful raids on Spanish ships or ports. They never found any fresh water, however, prompting their able commander, Woodes Rogers, to rail bitterly in his journal against earlier writers who "have given very blind or false relations" about water and timber on the Galápagos. "I shall say no more of these islands," he asserted, "since by what I saw of 'em, they don't at all answer the Description that those Men have given us."[44] Apparently Rogers's faith in Dampier suffered as well.[45]

Despite this curt dismissal of the archipelago, Rogers's published journal (a highly popular narrative) included two keen scientific observations about the place.[46] First, Rogers recognized the volcanic origins of its terrain and saw that this explained the absence of surface water. "The Island is nothing but loose Rocks, like Cynders, very rotten and heavy," he reported about one landfall, "and the Earth so parch'd, that it will not bear a Man, but breaks into Holes under his Feet, which makes me suppose there has been a Volcano there." Even where no volcano now exists, he reasoned, one could have erupted in the past and left a cinder-and-lava surface incapable of holding water.[47] Such an interpretation of the Galápagos terrain was far from obvious. Indeed, one of his chief officers, Edward Cooke, wrote about the same "Rocky, dry Ground" that it "looks as if there had lately been a Earthquake" rather than a volcano—but an earthquake, his commander realized, could never account for this burnt-over desolation.[48] Then, after noting the abundance of iguanas and land tortoises, Rogers raised a second critical scientific issue: "'Tis strange how the latter got here," he wondered, "because they can't come of themselves, and none of that sort are to be found on the [American] Main[land]."[49]

Both of these observations challenged traditional scientific conceptions. The assertion of ancient volcanos suggested development over time in geology. The comment about tortoises questioned received wisdom regarding the origin and distribution of land animals. In each respect, Rogers anticipated developments in scientific thought in which belief in a fixed creation gave way to notions of change over time.

Both Cooke and Rogers expressed contempt for the animals of the Galápagos Islands. Their tortoises, Rogers wrote, "are the ugliest in Nature, the Shell not unlike the Top of an old Hackney Coach, as black as Jet, and so is the outside Skin, but shrivel'd and very rough; . . . the Head little, and Visage small, like a snake, and look very old and black." Cooke wrote of Galápagos iguanas, "We shoot them among the Rocks, they are eaten by the Sea-men, and reckon'd good Meat, though their Deformity made me loath to eat them." Both writers scorned the local birds, which Rogers described as "so very tame that we often hit them down with Sticks." Most significantly, he added, "I saw no sort of Beasts," by which he meant land mammals.[50] European naturalists had long placed such "beasts" atop the created order in a great chain of being. Cold-blooded reptiles (including tortoise, turtle and iguana) typically ranked much lower. The absence of the higher and the predominance of lower reflected poorly on a place.

The abject fauna fit a larger pattern. European explorers and naturalists had questioned the relationship of New and Old World animals ever since their first encounter with America. Columbus, wondering about the lack of large land mammals, simply assumed he had not yet reached the Asian mainland. The wonder increased after further exploration confirmed America as a separate landmass with fewer types of large mammals than the connected continents of the Old World. Could both animal populations come from a single source and, if so, why did they differ? As long as each hemisphere contained representatives from all levels of the great chain of being, their relative distribution carried little significance. In the Americas, however, the sequence seemed incomplete, generating new applications of hierarchical thinking.

The concept emerged of an inferior America. Biological species either had never developed to their full potential in the New World environment, European naturalists reasoned, or they had degenerated there from their Old World vigor. As early as the 1500s, Oviedo had judged the New World too wet for mammals, with Acosta adding that this condition rendered the Americas largely uninhabitable by humans and higher beasts. At the time, many naturalists associated excessive wetness with the putrefaction of higher animals and the spontaneous generation of lower ones; and as went native America, so went native Americans: Bacon characterized them as "a young people . . . not like [the biblical] Noah and his sons," whose descendants peopled the Old World.[51] In a poetic reference to the newly discovered hemisphere, John Donne wrote of "That unripe side of earth, the heavy clime/ That gives us men up now, like *Adams* time."[52]

The emerging European view of American inferiority reached the height of contempt in the writing of the great eighteenth-century French naturalist Georges Louis Leclerc de Buffon, who held Old World quadrupeds in particularly high esteem and had nothing but disdain for New World reptiles. "Let us now see why there are found in the new world such large reptiles, such big insects, such small quadrupeds, and such cold men," he wrote in a typical passage. "This is accounted for by the quality of the earth, the condition of the sky, the degree of warmth and humidity, the situation, the elevation of the mountains, the quantity of running or stagnant water, the extent of the forest, and above all the crude state in which nature is found."[53] Most of these factors applied in the extreme to the Galápagos, where reptiles assumed monstrous proportions, insects flourished, mammalian quadrupeds scarcely existed and no humans had ever lived.

Beginning with Thomas Jefferson's *Notes on the State of Virginia* in 1784, some European-American naturalists defended the honor of their native species. These defenses, however, implicitly accepted Old World biological hierarchies and simply offered counterexamples of big American land mammals, such as bison, elk, bear and (in Jefferson's case) the fossil remains of a woolly mammoth from western Virginia. Such efforts failed to benefit the Galápagos Islands. As long as European hierarchies prevailed, the Galápagos, with its dominant reptiles, birds, insects and sea creatures, held little meaning for naturalists. Buffon's environmental explanations for the sorry state of the local fauna tended to depress interest still further by giving a naturalistic reason for the islands' seeming uselessness. Whether cursed by God or by nature, the archipelago was damned in pre-Darwinian European eyes.

TWO

THE NATURAL THEOLOGY
OF HELL

IN CROSSING A HEATH, suppose I pitched my foot against a *stone*, and were asked how the stone came to be there, I might possibly answer, that, for any thing I knew to the contrary, it had laid there forever." So wrote the sometimes Cambridge University lecturer and Anglican cleric William Paley to begin his profoundly influential 1802 treatise, *Natural Theology*. "But suppose that I had found a *watch* upon the ground, and it should be inquired how the watch happened to be in that place." Surely the former answer would not do. "The watch must have had a maker," Paley concluded, "who comprehended its construction, and designed its use."[1]

This most famous passage of English popular science writing resonated in its day because it echoed eighteenth-century developments in British scientific thought and articulated the view of nature widely held by scientists in Britain and the United States during the first half of the nineteenth century. Even though science had become a decidedly secular enterprise in parts of continental Europe by 1800, British and American scientists largely viewed nature through the lens of natural theology—expecting and finding evidence of God's existence and character in the world around them. Second only to God's revealed word in the Bible, Protestant natural theologians believed, God's physical creation told careful investigators about the divine. Natural theology thus offered a powerful reason to study science.

British and American thought set the tone for scientific understanding of the far-off Galápagos Islands because of the dominant role then played by Britons and Americans in studying the place. (Following general conventions of the time, Americans in this context and hereafter refers solely to people from the United States.) Increasingly during the eighteenth and nineteenth centuries, British and American ships ruled Pacific exploration and commerce—with the Galápagos Islands falling within that domain despite their proximity to South America. Spanish (and later Ecuadorian) sovereignty over the uninhabited archipelago meant virtually nothing.

Significant, though, was the gradual drift of English science from its heights in the 1600s to its depths in the 1700s. No single factor accounted for the slippage; but the Baconian utilitarianism that quickened science during the earlier period gradually gave way to the natural theology that dulled it during the latter one—a trend that spilled over into American science by the early 1800s.[2] The Galápagos Islands remained incomprehensible under both viewpoints, but natural theology at least prepared the way for future interest in the place when, in the late 1800s, William Paley's unchanging God gave way to Charles Darwin's evolutionary processes as the scientifically accepted designer of living organisms. Darwin in fact always credited Paley's writings with inspiring his own quest for a logical explanation of nature's functional orderliness, which led him to natural selection after he had rejected the theory of divine design.[3] Eighteenth-century naturalists, however, came first.

The religious tone of British natural history—a field of science that then broadly incorporated botany, zoology and geology—was fixed at the start of the eighteenth century with the publication of John Ray's *The Wisdom of God Manifest in the Works of Creation*. Before writing this classic work, the energetic Ray had laid the foundation for modern natural history in Britain by compiling the first systematic accounts of the kingdom's native plants, fish, mammals and reptiles. These early works reflected the Baconian utilitarianism that inspired the formation of the Royal Society, which Ray joined soon after its founding in the 1660s. "There are those who condemn the study of Experimental Philosophy as mere inquisitiveness," Ray had written in his *Synopsis of British Plants*, but "those who scorn and decry knowledge should remember that it is knowledge that makes us men, superior to the animals and lower than the angels."[4] Just this sort of practicality energized seventeenth-century British science: a conviction that God created nature for human uses that naturalists could readily discern through systematic observation and unbiased experimentation.

Ray, however, moved beyond Baconian natural history to full-fledged natural theology. Bacon, despite his professed Protestantism, had warned scientists against metaphysical speculation about God's purposes in the design of natural objects—better to stick with finding their uses. To say "that 'the clouds are for watering the earth'; or that 'the solidness of the earth is for the station and mansion of living creatures',," he cautioned naturalists, "is well inquired and collected in metaphysic, but in physic they are impertinent." Better to examine the structure of the clouds above or the terrain below (and exploit that knowledge) than to speculate about why they exist.[5]

With *The Wisdom of God*, Ray rushed in where Bacon feared to tread, and a century of British naturalists followed. This book, Ray wrote in its preface, would "serve not only to Demonstrate the being of a Deity, but also to illustrate some of his principal Attributes, as namely his Infinite Power and Wisdom" in the "admirable contrivance" of the earth and its beings. "How variously is the Surface of it distinguished into Hills and Valleys, and Plains, and high Mountains affording pleasant prospects? How becomingly clothed and adorned with the grateful verdure of Herbs and stately Trees?" Such pleasant musing about nature from pastoral England made no sense of the harsh Galápagos environment. Ray specifically portrayed clouds, water, plants, soil and mountains as so providentially distributed "as to render all the Earth habitable" by humans—yet even a casual visit to the Galápagos Islands (or any number of deserts, ice sheets or jungles) would give the lie to such anthropocentric assumptions.[6] If naturalists looked to nature for their theology, better to ignore places like the Galápagos—which most of them did.

Ray's brand of natural theology made matters even worse for understanding the Galápagos Islands by admitting no change in nature. "By the Works of Creation," he wrote, "I mean the Works created by God at first, and by Him conserved to this Day in the same State and Condition in which they were first made."[7] This definition either excluded the Galápagos from creation altogether or left their ongoing volcanic development inexplicable.

Creationist natural theology dominated British science for a century. The era became known as the Enlightenment in European thought, but while the Enlightenment on the continent (especially in France) grew increasingly secular, Britain stood apart. "Natural theology was the spiritual core of the British Enlightenment, . . . providing eager scientists with a convincing metaphysical justification for their research," one historian observed. "It functioned as the center of an intellectual consensus so pervasive, that it remained healthy and vigorous deep into the nineteenth century."[8] Among British naturalists of the era, another historian added, "everyone agreed that natural history must devote itself to exhibiting evidence of divine design and material Providence."[9] Yet the Galápagos Islands gave evidence of neither. Without any concept of geological or biological change, the archipelago was left all but outside the realm of natural history. It is a common pattern in science: observations not fitting into existing theories are often ignored.[10]

Lack of interest led most of the explorers who charted the Pacific during the eighteenth century to bypass the Galápagos. The magnificent voyages of James Cook are a case in point. He transformed the European journey of exploration from a haphazard affair into a carefully organized naval enterprise staffed by scientifically trained officers and professional naturalists.[11] Cook's three expeditions canvassed the Pacific for Britain from 1768 to 1780, looking for everything of scientific, economic or strategic value. They sailed right by the Galápagos without stopping.

Assigned to follow up on some of Cook's investigations, George Vancouver (who served on Cook's final voyage) sailed his great ship *Discovery* through the archipelago in 1795 with scarcely more than a puzzled nod. "The interior country exhibited the most shattered, broken, and confused landscape I ever beheld, seemingly as if formed of the mouths of innumerable craters of various heights and different sizes," Vancouver observed without disembarking. In his personal log, First Lieutenant Joseph Baker described Albemarle Island as "nothing but a large Cinder." The ship's naturalist, Archibald Menzies, who briefly went ashore on Albemarle with a small landing party, called it "the most dreary barren and desolate country I ever beheld." On shore, Menzies noted the curious mix of cold- and warm-weather beasts, jotting in his log: "Seals & Penguins in vast abundance, whilst the surf of the adjacent sea . . . swarmed with large Lizards." Vancouver pursued this point in his published account of the voyage, commenting that "these shores, however singular it may seem, abound with that description of those animals which are usually met with in the temperate zones, bordering on the arctic and antarctic circles." Vancouver also noted the "singularly temperate" climate of these equatorial islands and the strong tides swirling about them, but never connected his observations to the presence of a major cold-water current. After scant study of this "very dreary and unproductive" place, the *Discovery* sailed on.[12]

Expeditions from other European countries failed to pick up the slack. None of the French or Dutch explorers in the Pacific stopped at the Galápagos during the 1700s. After two centuries of official noninterest, Spain organized several expeditions to explore its New World holdings systematically with trained naturalists. Two of these briefly visited the Galápagos Islands; the first one led by Don Alonso de Torres in 1789 and the second and greater one led by Alessandro Malaspina a year later. No one ever published any reports from these visits, however, so their findings, if any, were lost to science. In 1799, the Spanish government authorized the German naturalist Alexander von Humboldt to conduct the explorations that opened Ecuador and much of South America to science, but he never crossed over to the Galápagos despite studying the oceanic current that cooled it. Although the archipelago constituted an unusually large island group by Pacific standards, and by far the largest anywhere near the American mainland, it was a useless anomaly to eighteenth-century European explorers and naturalists.

The Galápagos remained uncharted, except for crude buccaneer maps, and its plants and animals went uncollected long after Tahiti, Hawaii, Fiji and other major Pacific islands were comparatively well studied. When this began to change, early in the nineteenth century, it was not due to new scientific paradigms or a compulsion to fill blank spaces on navigational charts but rather because, for the first time since the decline of buccaneering more than a century earlier, British and American ships had begun venturing into Galápagos waters. They had a common goal: whales. The stories that whalers brought back to Britain and the United States rekindled interest in the place.

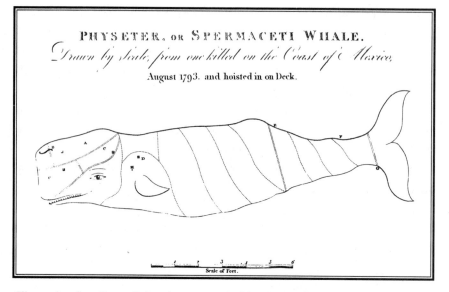

Illustration from James Colnett's report on Pacific whaling (1798)
(Source: Used by permission of UGA.)

The boom in South Seas whaling began and ended on technological innovations. Europeans primarily valued whales as a source of fuel oil, which they extracted by cooking the animal's fresh blubber. For centuries this limited the catch to coastal, cool-water right whales, whose blubber could be taken back to land for processing before it spoiled. As inshore stocks dwindled during the 1700s, ingenious New Englanders developed shipboard tryworks that allowed them to follow deep-water sperm whales into tropical waters and process their catch on the spot. Thus Herman Melville could soon write, "Of the grand order of folio Leviathans, the Sperm Whale and the Right Whale are by far the most noteworthy. They are the only whales regularly hunted by man."[13] Intrepid American and British whalers penetrated the Pacific in the 1790s and soon found a rich harvest among the Galápagos Islands, where the confluence of ocean currents produces ideal feeding and breeding grounds for whales. An oil rush ensued; it continued for over half a century, severely depleting stocks, until the discovery of commercial quantities of petroleum (beginning in Pennsylvania during the 1850s) and the means to process it provided a cheap substitute for whale oil.

For a few decades, however, whaling brought ships, sailors and scientists to the Galápagos. Not only had the islands gained a measure of economic value but, even seen through the perspective of natural theology, they and their strange animals had begun to make some sense to British and American observers. Torn between the archipelago's uninhabitability and its newly found beneficial uses, a series of early nineteenth-century observers struggled to come to terms with the place.

The first account of these efforts appeared near the turn of the century, written by the British navy captain James Colnett. With the coming of the war against France in the 1790s, Colnett sought command of a military ship; but as he had once served with distinction on Cook's second voyage in the Pacific, the admiralty, at the behest of influential London merchants, instead sent him in search of whaling grounds. Colnett began his investigation in the South Atlantic early in 1793, but by the following spring his quest carried him around the Horn and into Galápagos waters. "Here we cruised till the eighth of April, and saw spermaceti whales in great numbers," Colnett noted in his published journal. "I am disposed to believe that we were now at the general rendezvous of the spermaceti whales from the coasts of Mexico, Peru, and the Gulf of Panama, who come here to calve." In support of this conclusion, Colnett reported observing many of these whales "in a state of copulation," an impressive act that neither he nor any of "the oldest whale-fishers, with whom I have conversed" had ever before witnessed. Here was the mother lode of whales. "I frequently observed the whales leave these isles and go the Westward, and in a few days, return with augmented numbers," he wrote ecstatically. "I have also seen the whales coming, as it were, from the main, and passing along from the dawn of day to night, in one extended line, as if they were in haste to reach the Galipagoes."[14]

Colnett and his crew visited the archipelago twice during their two-year cruise. Their eyes now open to its animal life, they beheld its marvels—perhaps the first people to appreciate them so fully. "In this expedition we saw great numbers of penguins, and three or four hundred seals," Colnett wrote of their stop on one island. "There were also small birds, with a red breast, such as I have seen at the New Hebrides; and others resembling the Java sparrow, in shape and size, but of a black plumage; the male was the darkest, and had a very delightful note." Clearly struggling to relate the local animals to known types from other places, Colnett repeatedly resorted to the superlative. "There is great plenty of every kind of fish that inhabit the tropical Latitudes; mullet, devil-fish, and green turtle were in great abundance," he noted. "But all the luxuries of the sea, yielded to that which the island afforded us in the land tortoise, which in whatever way it was dressed, was considered by all of us as the most delicious food we had ever tasted. The fat of these animals when melted down, was equal to fresh butter." From feasting on these great beasts, Colnett exalted, "all apprehensions of scurvy or any other disease was at an end." By the time they departed the Galápagos, "everyone was charmed with the place."[15]

In an age that sought God's divine design and providence everywhere in nature, Colnett began to see them on the Galápagos as well. Of course a breeding ground for whales stocked with delicious and nutritious tortoises represented a Godsend. The place even offered respite for sailors from the oppressive tropical heat. "In the morning, evening and night, it was below summer heat in England," Colnett marveled. "I consider it as one of the most delightful climates under heaven." Even the lack of fresh water seemed tempered by an abundance of succulent plants. Tortoises sucked water from tree bark, birds drew it from leaves, and in one touching scene of benevolence in nature, sailors "observed an old bird in the act of supplying three young ones with drink, by squeezing the berry of a tree into their mouths."[16] Such complex interactions seemed certain evidence of divine beneficence.

The inhospitality of the islands ultimately darkened Colnett's vision of them. His second stop at the archipelago, nine months later, apparently coincided with one of its periodic droughts. Rock cavities and hollows that once retained rainwater for human and animal consumption were now dry. The succulents had withered. "An officer and party, whom I sent to travel inland, saw many spots, which had very lately contained fresh water, about which, the land tortoises appeared to be pining in great numbers," Colnett noted. As for the small land birds once seen resourcefully squeezing berries and piercing leaves for water, "on our return, we found great numbers dead in their nests, and some of them almost fledged."[17] He was particularly struck by the sight of an Albemarle Island bay once frequented by buccaneers. "The inhospitable appearance of this place was such as I had never before seen, nor had I ever beheld such wild clusters of hillocks, in such strange irregular shapes and forms, as the shore presented, except on the fields of ice near the South Pole." Colnett's note in the ship's log for the day is more stark: "Saw nothing but an Iron Bount Inhospitable barren coast."[18]

The volcanic terrain was a perennial blight on the land. "In some parts, it has the appearance of being covered with cinders; and in others, with a kind of iron clinker," Colnett noted. Neither type of terrain supported a rich flora or fauna. "There is no tree, in this island, which measures more than twelve inches in circumference," he commented early on, later adding that the dwarfed vegetation was "not large enough for any purpose, but to use as fire-wood." At another point, a landing party returned with a typical report: "They saw no esculent vegetable, nor found any water that was sufficiently palatable to drink." Still later Colnett lamented, "No quadruped was seen on this island, and the greatest part of its inhabitants appeared to be of the reptile kind." They found the "native ugliness" of the land iguanas "so disgusting . . . that no one on board could be prevailed on, to take them as food." Even worse, "There were, besides, several species of insects, as ants, moths, and common flies, in great numbers."[19]

Struggling to come to terms with the place, Colnett ultimately concluded that, at least for whalers, the bounty of the Galápagos's sea life outweighed the bane of its landscape. Perhaps that made the archipelago cognizable even within the constraints of natural theology. Situated amid breeding grounds for whales, with tortoises to eat and enough seals to "form no inconsiderable addition to the profits of a voyage," Colnett opined, "these isles deserve the attention of British navigators."[20] They received that attention for as long as the South Seas whaling boom continued.

To assist these future navigators, Colnett's report included a large chart of the archipelago that combined Cowley's earlier work with his own new reckonings but still omitted many islands, including several major ones. The new chart retained British buccaneer names for most of the known islands—James, Charles, Albemarle, Narborough—but recognized a later generation of British officials in the naming of newly found or relocated ones—including Chatham, Barrington, Duncan and Hood.[21] Although the early twentieth-century Galápagos travel writer Victor von Hagen hailed this as "the first modern map of the Galápagos Islands," he never

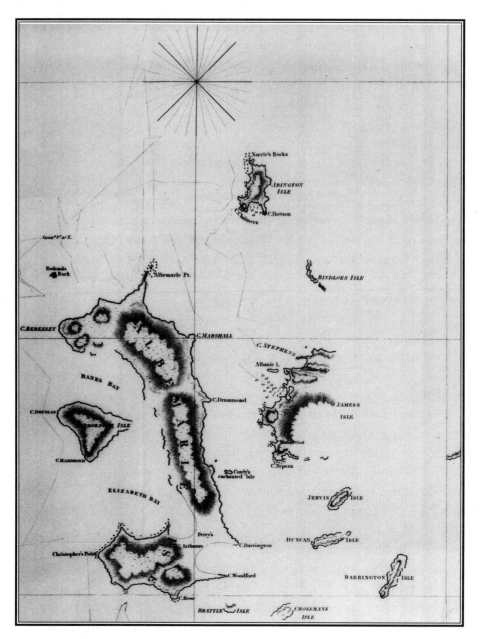

Colnett's chart of the Galápagos Islands (1798)
(Source: Used by permission of UGA.)

had to rely on it.[22] Those who did complained bitterly about its errors. Yet still they came, drawn by Colnett's promise of whales, seals and tortoises. The later Galápagos explorer Joseph Slevin could fairly class Colnett's expedition as "no doubt the most outstanding of the early voyages" to the archipelago.[23]

For the next half century, whalers flocked to Galápagos waters. Many came from Great Britain, as Colnett and his sponsors had hoped, but even more from New England. Few ships hailed from anywhere else, except for the occasional Spanish raider harassing the foreign whalers. "Rounding the storm-tossed Horn, where ship and man were beaten by the tantalizing wind," von Hagen wrote, "whaling ships entered the halcyon Southern Sea and followed the cold Humboldt Current northward along the coast of Chile and Peru, a current richly red with plankton. . . . After cruising northward, the whalers, following a time-honored course, then turned due westward at the equinoctial line and made for the Galápagos Islands."[24] Once there, these whalers typically cruised the vicinity for months, regularly encountering other whalers and exchanging information between crews. The sailors even erected a "whalers' post office" on Charles Island—an upturned barrel in which persons from one ship could leave messages for those on another, and anyone could post letters for homebound vessels to carry back with them.

Rivalries sometimes developed between American and British sailors during those years of heightened nationalism between the American Revolution and the War of 1812. Although both sides relied on Colnett's report and chart, the Americans criticized them the most—perhaps to diminish any claim to the place that Colnett's expedition might give to the British. Captain Amasa Delano, who came from the same Massachusetts family that later produced Franklin Delano Roosevelt, exemplified the American side. Delano's published description of his voyage to the Galápagos opens with a jab at Colnett's description of the anchorage at Stephen's Bay, on Chatham Island. "I presume when captain Colnett recommended it for vessels of two hundred tons," Delano wrote, "he had not sounded round its mouth, or within the cove; if he had, he would have found that there were many places which have not more than six feet water, and a very rough ledge of rocky bottom."[25] Delano, who as a boy had volunteered to fight the British in the American Revolution and never stopped sniping at them thereafter, filled his journals with sneers and slights against Colnett and all things British.

Delano visited the Galápagos Islands in 1800 under unusual circumstances. In command of a merchant vessel engaged in the opium-lubricated China trade, he stopped there to allow his crew time to recover from scurvy. This gave the able-bodied Delano ample opportunity to explore the surroundings, and his published journal offered some of the best early descriptions of Galápagos reptiles, including the first known report of the lava lizard (noting its "bright vermilion red throat"). The islands' giant tortoises received the most attention, of course: Delano's detailed description of their physical characteristics, behavior and diet extends for pages. "I have seen them with necks between two and three feet long," he noted, "and when they saw any thing that was new to them, or met each other, they would raise their heads as high as they could, their necks being nearly vertical, and advance with their mouths wide open; appearing to be the most spiteful of any reptile whatever. Sometimes two of them would come up to each other in that manner, so near as almost to touch, and stand in that position for two or three minutes." At the slightest touch, however, they would immediately retreat into their shells in fright. "They are perfectly harmless," Delano added, eating "grass, or any flowers, berries, or shrubbery." Based on his shipboard observation of captured tortoises, Delano was the first to suggest that they survived long periods of drought in the harsh Galápagos environment by consuming stored fat.[26]

Turning to native birds, Delano contributed detailed accounts of the brown pelican, the Galápagos dove and "another remarkable bird found here, which has not before been described" that for obvious reasons became known as the blue-footed booby. "They resemble the small kind of booby," he observed, "excepting they are of rather a darker colour on the breast and neck, and their beaks and feet are of a Prussian blue." Delano marveled at their ability to dive for food "from sixty to a hundred yards in the air," adding that they "go into the water with the greatest velocity that can be conceived of, exceeding any thing of the kind that I ever witnessed." Plunging to great depths through the transparent sea, "they glide under water at almost as great a degree of swiftness as when flying in the air."[27] In an era when Anglo-American science was dominated by amateur naturalists writing for educated readers generally, these observations stand out for their descriptive precision.

As in Colnett's journal, however, gloomy reports of the land tempered the glowing accounts of the animals, leaving a hellish overall image. Delano described Chatham Island as "mountains of rocks burnt to a cinder." On Hood Island, he wrote, "the surface is burnt stones and sand, with some small shrubby wood growing on it." Albemarle's "appearance is like most of the other large islands, a great part of it appearing to have been torn to pieces by volcanos." As if to punctuate the burnt-over aspect of the place, Delano watched one of Albemarle's volcanos erupt: "the most extraordinary phenomena . . . that I have ever witnessed in my life." Drought added to the desolation, leading Delano to comment "that it never rained at these islands" and to write them off for human habitation.[28] Melville carried Delano's grim depiction of the Galápagos landscape into his sketches of the Enchanted Isles, just as he drew on Delano's gruesome account of saving a Spanish captain from slave-ship mutineers in crafting "Benito Cereno."[29]

Melville drew even greater inspiration for his sketches from the next notable island visitor, American navy hero David Porter. At the outbreak of its war with Britain in 1812, the United States government despatched Porter, in command of the frigate *Essex*, to protect American whalers in the South Seas and harass British ships there. This brought him to the Galápagos Islands, where he cruised from April to September, 1813. The *Essex*'s total success in its mission, which forced the British admiralty to send a small fleet in pursuit, made Porter's name second only to John Paul Jones in early American navy lore.[30] Acclaimed in the United States and denounced in England for its tales of the *Essex*'s prowess, but widely read in both places, Porter's *Journal of a Cruise Made to the Pacific Ocean* never went out of print. Porter's work became one of the most significant travel books for awakening readers to the distinctive natural history of the Galápagos Islands, second only to Darwin's *Journal of Researches*.

Following Colnett, whose published journal and chart he took on the voyage, Porter condemned the Galápagos as uninhabitable. "These islands are all evidently of volcanic production; every mountain and hill is the creation of an extinguished volcano," he observed. "Thousands of smaller fissures, which have burst from their sides, give them the most dreary, desolate, and inhospitable appearance imaginable. The description of one island will answer for all I have yet seen; they appear unsuited for the residence of man."[31]

Porter found the islands even more forbidding than Colnett had described them because, like Delano, he arrived during a severe drought that browned the scant vegetation and dried the freshwater sources identified in Colnett's journal. Provoked perhaps by nationalism and thirst, both American authors repeatedly scored Colnett on this point. "We find captain Colnett's chart of the island, so far as he surveyed it, sufficiently accurate for our purposes," Porter sneered in a typical passage, "but we neither found his delightful groves [nor] his rivulets of water." Porter's wit resurfaced when he described the route to one of Colnett's then-dry water holes. "You may proceed for about three miles without experiencing much inconvenience, except from the intense heat of the sun, and from occasionally falling into the holes made by the guanas in the loose cinders, heated by the sun's rays, as well as from occasionally encountering in your route beds of sharp lava, about as agreeable to walk on as hackle," he observed. "This transition from hot to sharp and from sharp to hot is equally desirable, for either of the evils is so great that it cannot be long borne, and of the two it is difficult to say which is the least." By the time the *Essex* and its captured ships sailed from the Galápagos, the officers and crew "were heartily tired of being confined to this most desolate and dreary place, where the only sounds to be heard were the screeching of the seafowls, and the melancholy howlings of the seals."[32] They could imagine no place more alien.

Unlike his predecessors, however, Porter did not view the islands' condition as permanent. In the years since Colnett's voyage, the Scottish gentleman naturalist James Hutton had completed his revolutionary *Theory of the Earth*, which helped transform how geologists understood the development of the earth's features. Hutton's work built on a generation of fieldwork, mostly by French naturalists, that finally began to recognize the widespread occurrence of volcanic activity and its impact in shaping terrain. Hutton compiled these scattered insights into a comprehensive geological theory, popularized during the early nineteenth century by the Scottish science writer John Playfair.[33]

Hutton envisioned a cyclical process of terrestrial creation and deterioration quite unlike earlier notions of geological origins inspired by the Book of Genesis. Simply stated, he postulated that subterranean pressure from the weight of sedimentary rock, coupled with the earth's internal heat melting the lower strata of that rock, pushed up volcanic mountains from beneath the ocean's floor. These then weathered back down, creating more sedimentary rock and beginning the cycle over again. According to Hutton, land became inhabitable by humans during the period after it had eroded enough to create soil but before it resubmerged under the sea. The earth thus displayed "no vestige of a beginning—no prospect of an end." Although this radical conclusion aroused opposition from biblically orthodox natural theologians, it stimulated widespread discussion in Britain and the United States during Porter's day. Traditional geological catastrophism (which presumed one or more worldwide acts of creation followed by gradual deterioration) remained ascendant for another generation, but Hutton's theory made better sense of the widespread evidence of past and present volcanic activity found by European explorers and naturalists.[34] It made perfect sense of the Galápagos.

Porter never mentioned Hutton by name, but Hutton's thinking ran through his interpretation of the Galápagos. As the most volcanically active, the western islands were the newest, Porter reasoned: vulcanism must still be pushing them up from below the sea floor. Narborough Island, which erupted during his visit, "probably owes its origin to no distant period"; its "hills composed of ashes and lava, all apparently fresh, and in most parts destitute of verdure, sufficiently prove that they have not long been thrown from the bowels of the ocean." Returning to the island after its eruption, Porter added that "Narborough appeared to have undergone great changes since our last visit." Four of its separate craters had begun spewing forth smoke, as did one crater on nearby Albemarle, suggesting "a submarine communication between them."[35] Here were ongoing, interconnected geological forces sufficient to shape the earth's features, just as Hutton's theory predicted: no need for an initial creative act to do the sculpting.

Porter noted subtle differences as he moved from west to east within the archipelago. He wrote of Chatham, in the extreme east, "This island, like all the rest, is of volcanic origin, but the ravages appear less recent." Speaking of it and two central ones, James and Charles, he added, "The soil of these islands, although dry and parched up [due to drought], seems rich and productive; and, were it not for the want of streams of fresh water, they might be rendered of great importance to any commercial nation that would establish a colony on them." Thus soil had appeared with time, again in accord with Hutton's theory. Given more time, Porter suggested, the light volcanic soil would compact sufficiently to hold water, permitting "springs or streams of water, for the support of animal life." Although all of the islands remained uninhabitable by humans, he told of four goats from his ship escaping onto James Island and apparently surviving in its moist highlands. Porter also related the story of a marooned sailor named Patrick Watkins who survived for years on Charles Island, and managed to grow some vegetables there. "We have seen, from what Pat has effected, that potatoes, pumpkins, &c., may be raised," Porter commented, "and with proper industry the state of these islands might be much improved." They remained too new for now, he acknowledged, but "time, no doubt, will order it otherwise; and many centuries hence may see the Gallapagos as thickly inhabited by the human species as any other part of the world."[36]

Such notions of change over time in Galápagos geology did not lead Porter to such thinking in biology. Neither did he explain the archipelago's distinctive animal life by the traditional biblical concept of a single creation. Rather he adapted a popular new idea developed by the great French naturalist Georges Cuvier during the early 1800s. From his careful study of the fossil record, Cuvier established to the satisfaction of nearly all naturalists that various animal species appeared and became extinct over a long span of geological time. Seeing no evidence of evolutionary development in fossilized species—even though they grudgingly looked for it—Cuvier's followers postulated a series of creation events. Some believed that these came from the hand of an ever-active transcendent God, while others attributed them to vital forces within nature.[37] Porter favored the latter alternative. "I shall leave others to account for the manner in which all those islands obtained their supply of tortoises and guanas, and other animals of the reptile kind," he wrote. "I shall merely state, that those islands have every appearance of being newly created, and that those per-

Gallapagos turtle

Galápagos Land Tortoise, from David Porter, Journal of a Cruise *(1822)*
(Source: Used by permission of UGA.)

haps are the only part of the animal creation that could subsist on them." As to how they got there if they could not fly or swim across the open ocean, Porter simply asked, "Nature has created them elsewhere, and why could she not do it as well at those islands?"[38]

The possibility of continuous separate creations of species suited to their locale led Porter to see various novel species scattered throughout the archipelago. Citing physical differences among tortoises, for example, he observed that "those of James' Island appear to be a species entirely distinct from those of Hood's and Charles' Islands." Galápagos Tortoises as a group also stood apart from all other kinds, he noted, and "properly deserve the name of elephant tortoise" due to their enormous size.[39]

Porter found subtle differences among other animals as well. Sea turtles in Galápagos waters "were shaped much like the green turtle, but were of a black, disgusting appearance." Later naturalists would call them a subspecies. He identified some land birds as "peculiar to those islands," and remarked at their tameness. This, he wrote, "offered great amusement to the younger part of the crew in killing them with sticks and stones."[40] By the time he left the archipelago, Porter far surpassed Colnett in seeing the dynamic character of geology and biology on the Galápagos—and reveled in taunting him from the pages of his journal.[41]

The end of the War of 1812 brought true peace between British and American ships on the high seas for the first time since the revolution. Whalers and sealers from both countries plied Galápagos waters, stimulating interest in the archipelago and providing an opportunity and economic rationale for further scientific study of the place.

By midcentury, the United States boasted a 700-ship whaling fleet and somewhat fewer sealers, mostly based in New England but many operating in the Pacific. After rounding Cape Horn, countless American ships stopped in the Galápagos Islands to stock up on tortoises and cruise for whales or hunt seals in the vicinity. Captain Benjamin Morrell, for example, in his *Narrative of Four Voyages to the South Seas*, wrote of harvesting 5,000 fur sealskins on the Galápagos in 1823, witnessing a volcanic eruption of Narborough Island early in 1825 and later that year taking 187 tortoises from an island he called Indefatigable. "These islands are all of volcanic origin; and have, generally speaking, always been barren," he commented, citing Porter as his source. "But of late years they have become more fertile, both the upland and valley being now tolerably well wooded, over a good and rich soil, which wants nothing but a more liberal supply of moisture."[42] Other sailing logs from the period talk of individual ships taking up to 900 tortoises from the Galápagos—all for the sailors' mess—until the animals were nearly extinct in coastal areas by midcentury.[43] Once taken, the tortoises could live for months without food or water, until killed for fresh meat.

The other major published account of an American visit to the archipelago between the time that Porter left in 1813 and Darwin arrived in 1835 came from J. N. Reynolds, private secretary to U.S. Navy commodore John Downes on the frigate *Potomac*. In 1831, the United States sent the *Potomac* under Downes to the East Indies island of Sumatra to punish a group of Malayans who had seized an American merchant ship and killed its crew. On the return voyage in 1833, Downes (who had cruised Galápagos waters two decades earlier on the *Essex*) stopped by the archipelago for supplies and to show support for American whalers in the region.[44] He found it had changed since his first visit.

A year earlier, the newly independent nation of Ecuador laid claim to Galápagos by establishing a small settlement on Charles Island, which it renamed La Floriana. The colony prospered for a time under the leadership of Governor José Villamil, a native of New Orleans who moved to Ecuador following the Louisiana Purchase and rose to the rank of general there during the wars of South American liberation from Spain. Although crops failed near the barren coast, the settlers soon found that they could raise vegetables and livestock in the interior highlands, at least during the wet season. "There is a beautiful upland valley," Reynolds noted. "The soil is of a superior mould . . . composed of the decomposition of lava and vegetable matter"; it wanted only water in the dry season. When he visited the colony less that one year after its establishment, Reynolds found settlers doing a brisk business selling surplus produce and meat to passing ships. Thirty-one whalers had stopped at La Floriana during the period, he reported, all but two of them from the United States. Reynolds expressed great optimism for the small settlement, but it soon disintegrated after Ecuador began exiling criminals there, driving off the original settlers.[45]

American navy ships of the day, even those on long voyages to remote shores, rarely carried trained scientists or supported research. (The U.S. Navy's one magnificent exception, the expedition led by Captain Charles Wilkes from 1838 to 1842, bypassed the Galápagos during its several transits of the Pacific.) The *Potomac* fit the American norm. Its four-year-long circumnavigation netted little for science and practically nothing from the Galápagos. Upon his return, Reynolds deposited fifteen specimens of Galápagos plants with the Boston Society of Natural History together with one sheet of colored drawings of Galápagos fish. Somewhat more striking, the society's *Journal* reported the receipt from Commodore Downes of "two gigantic Galápagos Tortoises (living) weighing near three hundred and twenty pounds each." The great beasts survived as a feature attraction at the society for years.[46]

Reynolds and Downes's haphazard collection of exotic plants and animals reflected a growing obsession within Western science: categorizing everything in nature. Until modern times, species' names often differed from place to place, even within Europe, and displayed little overarching order. Premodern European naturalists classified the living world according to conflicting characteristics—one book might lump together yellow birds, for example, while another grouped small ones. This haphazard approach sufficed so long as naturalists had to cope with only a few hundred common European species, but as explorers brought back thousands of new species from other regions, it grew unwieldy. Without a common system of nomenclature, naturalists could neither distinguish novel species from previously reported ones nor make sense of the whole.

In the mid-1700s, Swedish naturalist Carolus Linnaeus solved this problem by developing a uniform system of binomial nomenclature based on resemblances in a single characteristic. For higher plants, for example, he chose flower parts. Plant species with nearly similar flowers formed a genus; genera displaying somewhat similar flowers comprised an order; orders with grossly similar flowers made up a class. "The success of Linnaeus' artificial system depended on the ease of assigning any species to its correct position," historian of biology Peter J. Bowler explains. "To discover the class and order of a particular plant, it was necessary only to count the stamens and pistils of its flower. Orders were divided into genera and species by closer inspection, taking into account the shape and proportion of the flowers."[47] Other simple rules applied for classifying animals and nonflowering plants. Each species became known to science by its unique genus and species name—assigned in Latin by the first naturalist to identify it according to the new system. Never mind that the plant or animal already had a local name, European scientists now would not recognize it without a proper Latin one. Thus the Galápagos Tortoise became (from Porter) *Geochelone elephantopus*, and the blue-footed booby, *Sula nebouxi*.

Linnaeus began by classifying the plants and animals around him, and his followers did the same in their locales. This left a world of unnamed species outside Europe. Collecting and classifying them became something of an obsession with professional and amateur naturalists, especially among the British. Just as certain people cannot put down a crossword puzzle until every square is filled, so these naturalists seemed driven to classify every species once a system existed for doing it. Scientific organizations and scientifically inclined individuals competed in collecting exotic species in the form of dried herbarium specimens, animal skins, skeletons, fossils and garden or zoo items. Making field notes as Oviedo or Dampier had done no longer sufficed. Aspiring young naturalists volunteered for arduous voyages of exploration to remote parts of the earth, while wealthy patrons of science hired collectors to gather specimens on their behalf.

Of course, British naturalists could only go where British ships went. Few could afford to charter their own vessels. For biologically rich yet largely unexplored places in the early 1800s, this increasingly meant South Sea islands, where whaling and trading drew innumerable British ships. Practical opportunity finally coincided with scientific interest to begin opening the Galápagos to British science.

Discovery's naturalist, Archibald Menzies, made the first proper scientific collection of Galápagos species in 1795. He gathered "a few" plant specimens during a cursory afternoon visit to Albemarle Island as part of a landing party sent ashore by Captain Vancouver to look for food and fuel. Although these specimens made their way to the British Museum, they became mixed with *Discovery* specimens from Hawaii, so that Menzies lost priority for these finds. The rugged lava-bed shrub *Scalesia menziesii*, for example, became *Scalesia affinis*. For his part, Vancouver regretted the lack of time to conduct a navigational survey of the islands and to examine the "uncommon" local birds, tasks he urged on future expeditions.[48]

Published journals from two subsequent British navy expeditions to the region generated further interest in the Galápagos Islands as a place for scientific study. The frigates *Briton* and *Tagus* cruised through the archipelago during the War of 1812 in pursuit of the *Essex*, spotting the goats that had escaped from Porter's ships and stocking up on tortoises.[49] In his published account of the adventure, Lieutenant John Shillibeer of the *Briton* said of Albemarle Island, "It possesses no fresh water, but the numerous plants and shrubs would, to a botanist, be a source of infinite gratification. Many of those plants, which are exceedingly beautiful, grow immediately from solid lumps of black lava."[50] On a regular tour of duty supporting British interests throughout the Pacific in 1822, Captain Basil Hall anchored his warship at Abingdon Island long enough to conduct a scientific test of the earth's compression at the equator—but not long enough to do much else. "We had no time to survey these islands," he wrote, "a service much required, since few if any of them are yet properly laid down on our charts."[51]

Next came the most renowned British botanical collector of the era, David Douglas, but again with too little time to accomplish much on the Galápagos. The Horticultural Society of London, then actively engaged in obtaining exotic plants for British gardens, booked passage for Douglas on the Hudson Bay Company's brig *William and Anne* on its 1824 voyage to investigate company claims in the Oregon country. Wet, cool Oregon should contain many plants suitable for English gardens, society officials guessed. Douglas proved them right by introducing more than 200 plant species from the Pacific Northwest to Europe, including his most famous namesake, the towering Douglas Fir.[52]

En route, the *William and Anne* paused in the Galápagos to take on supplies for three days in January 1825, giving Douglas an opportunity to go ashore on James Island for two hours each day. He made the most of this opportunity, collecting specimens of 175 plant species and a scattering of seeds. "The whole of the Gallipagos are mountainous and volcanic," Douglas wrote in his journal. "Their verdure is scanty, as compared with most tropical counties, owing, apparently, to the parched nature of the soil and the absence of springs of fresh water." He noted a few large trees in valleys but little variety of species. "The birds, however, are abundant, and some of them exceedingly handsome. . . . I killed forty-five individuals of nineteen genera, all of which I skinned carefully, and had the mortification of losing all but one, a species of *Sula*, from the constant rain that prevailed for twelve days after leaving the Gallipagos."[53] The same rains destroyed all but forty of his plant specimens as well.

Ship's surgeon John Scouler accompanied Douglas on two of his shore visits to James Island. Scouler had studied under the influential botanist William Hooker at the University of Glasgow, and Hooker had secured for his former student the medical post on the *William and Anne* with the hope that Scouler would collect plant specimens along the way. Scouler did not disappoint, returning with a wealth of specimens from Oregon but few from the Galápagos.[54] "On penetrating into the country [on James Island] we found very few plants, at least few in comparison to what one might expect in such a climate," Scouler commented in his journal. "Indeed, the heat & moisture of the country was so great as to prevent us from preparing even the few that we thought were nondescript [or unknown]."[55] Yet thirteen Galápagos plant species first collected by Scouler eventually made it into the hands of William Hooker's son and successor as director of England's Royal Botanic Gardens at Kew, Joseph Hooker. The younger Hooker included them among the 236 species named in his 1847 publication, "Enumeration of the Plants of the Galápagos Archipelago," a work principally devoted to reporting on collections from the voyage of the *Beagle*. Hooker's paper also described five plants collected by Douglas.[56]

Scouler's disappointment with the Galápagos flora did not turn him against the islands. "The abundance & interesting nature of the animals amply compensated for the scarcity of the plants," he observed, noting in particular the Galápagos Tortoise, land iguana and a species of *Sula* with "very bright azure blue" feet. In addition, "To one whose knowledge of rocks & of geological phenomena had been confined to the primitive & transition districts of Scotland, James Island presents a new series of geological phenomena of the utmost interest." He closed with a call for further study of the place.[57]

The Galápagos Islands had become such a popular way station for Pacific travel that this call received an answer long before it was published. Scarcely two months after the *William and Anne* sailed out of Galápagos waters, the *Blonde* sailed into them under the command of George Anson, Lord Byron. Having recently succeeded to the title of his deceased nephew the poet, the new Lord Byron—already a navy captain and never expecting the peerage—received the assignment to transport home the bodies of the king and queen of Hawaii, who had died on a state visit to England. En route to Hawaii, the *Blonde* stopped at Albemarle Island for food and fuel. "The place is like a new creation," Byron wrote, "the birds and beasts do not get out of our way; the pelicans and sea-lions look in our faces as if we had no right to intrude on their solitude; the small birds are so tame that they hop upon our feet; and all this amidst volcanoes which are burning around us on either hand." After gathering ample firewood and four dozen sea turtles, the *Blonde* sailed on. "Our botanist found several rare and interesting plants, some of which are probably quite new," Byron commented, "however, our stay here was too short to procure any thing like a perfect catalogue of the natural productions of the Islands."[58] During his brief visit, ship's botanist James McRae collected forty-one plant specimens, which he turned over to the Horticultural Society of London. Thirty-seven of these made it into Hooker's 1847 "Enumeration"; twenty of them represented species new to science and another fourteen had never before been recorded from the archipelago.[59]

One final collector visited the Galápagos before Darwin, demonstrating just how easy it had become to reach the once-remote archipelago. A British expatriate and avid shell collector named Hugh Cuming, then living in Chile, sailed a hand-built vessel to the islands in 1829, where he gathered a boatload of shells and at least ten plant specimens. Upon his return to Great Britian, he sold the lot for display at the 1832 exhibition of the Zoological Society of London. Five of Cuming's plant specimens subsequently made Hooker's "Enumeration."[60] Clearly Galápagos specimens had become something of a scientific commodity by the 1830s. British navy captains Vancouver, Hall and Byron called for definitive navigational charts of the archipelago, and such experienced collectors as Douglas, Scouler and McRae spoke of the unknown species still there for the taking. Colnett, Scouler and Porter wrote of geological wonders demanding the attention of science. The Hookers, father and son, lobbied from within the British scientific establishment for more study of the archipelago. The British government responded by placing the Galápagos on the itinerary of its next Pacific expeditionary ship, HMS *Beagle*.

Obsessed with classifying everything in nature and bored with the commonplace at home, European naturalists sought novelties abroad. Yet the exotic biology and geology of places like the Galápagos often appeared in the context of an extreme environment. Later evolutionists would say that harsh conditions generate biological diversity. At the time, however, the fascination with collecting and naming drew attention to just those places that least fit prevailing notions of divine design.

In his classic treatise on natural theology, William Paley had maintained that beneficence in nature demonstrated the goodness of its Creator. Equating goodness with happiness and suffering with evil, Paley proclaimed, in a passage absolutely essential to his argument, "It is a happy world after all. The air, the earth, the water, teem with delighted existence. In a spring noon, or a summer evening, on whatever side I turn my eyes, myriads of happy beings crowd upon my view."[61] What then of the equatorial Galápagos, where a distinct spring never came and summer simply meant drought? "When the dews fail in the summer season, thousands of these creatures perish," Colnett wrote of the small Galápagos land birds. Another visitor wrote of a summer day on the archipelago in 1818, "The Birds and Snakes are frequently found dead in numbers during the dry season, and are supposed to perish from want of water." Porter blamed such deaths on the volcanic soil. "These thirsty mountains, like a sponge, soak from the clouds the moisture," he explained, "but they permit none of it to escape in springs or streams of water, for the support of animal life."[62] The volcanic, clinkerbound and forlorn Galápagos Islands appeared far removed from Paley's happy world, whose divine design produced joyful creatures in a beneficent creation. Yet commerce, empire-building and competition among naturalists meant that this anomalous place would no longer be ignored. The Galápagos Islands were about to play their role in shaping the modern mind.

THREE

WHAT DARWIN SAW

THE *BEAGLE* REACHED THE GALÁPAGOS ISLANDS on September 16, 1835, anchoring first near the northwest end of Chatham Island. "We landed upon black, dismal-looking heaps of broken lava, forming a shore fit for Pandemonium," Captain Robert Fitzroy wrote in his *Narrative of the Surveying Voyages of His Majesty's Ships Adventure and Beagle*.[1] His companion, the young naturalist Charles Darwin, had a similar reaction. "Nothing could be less inviting than the first appearance. A broken field of black basaltic lava is every where covered by a stunted brushwood, which shows little signs of life," he noted in his published *Journal of Researches*. "Although I diligently tried to collect as many plants as possible, I succeeded in getting only ten kinds; and such wretched-looking little weeds would have better become an arctic, than an equatorial Flora."[2] In his personal diary, which served as the basis for his published account, Darwin added here, "The country was compared to what we might image the cultivated parts of the Infernal regions to be."[3] At first blush, Darwin saw the place much as his predecessors had. This perspective reflected his past; it did not reveal the future.

Although Darwin filled the niche of naturalist on the *Beagle*, that small ship and its pedestrian mission did not merit an official scientist. Even after the *Beagle* returned to England in 1836, the president of the Royal Geographical Society patronizingly described Darwin as a "zealous volunteer."[4] Unlike the great voyages of Cook and Vancouver, who set out in tall ships staffed by experienced scientists to discover new lands for Britain, this trip, under the command of the twenty-six-year-old Fitzroy on a ninety-foot converted coastal carrier, simply aimed at completing a survey of South America's southern shoreline. By 1820 the region had broken free from Spain and opened its markets to non-Spanish trade for the first time, and British merchants and investors needed a harbor and coast survey to capitalize on the opportunity. They also sought safe new sea routes to the Pacific through desolate Tierra del Fuego to replace old ones around treacherous Cape Horn.

A two-vessel British expedition began the survey in 1826 under the overall command of Philip Parker King. The melancholic Pringle Stokes captained the expedition's smaller ship, HMS *Beagle*, used to explore narrow straits and shallow harbors. Lonely in command and lost in the stormy labyrinth of Tierra del Fuego, Stokes blew out his brains with a pistol in his poop deck cabin. "The soul of a man dies in him," he wrote in the logbook.[5] The British Admiralty sent the wellborn and well-trained junior officer, Robert Fitzroy, to take charge of the *Beagle* but King and both ships returned to England in 1830 without finishing the survey. A year later the *Beagle*, still under Fitzroy, was sent back alone to finish the job—a projected two-year task.

Enthusiastic about the prospect of command at such a young age and technically competent in surveying, Fitzroy nevertheless feared the psychological toll of such a long voyage without a suitable traveling companion. For Fitzroy, a direct descendant of King Charles II by the mistress Barbara Villiers and born of wealthy aristocrats on both sides of his family, this meant an educated, well-bred gentleman about his own age. Fitzroy deeply felt the need for such companionship because he exhibited a similar temperament as Stokes and suffered a similar fate. His uncle, Lord Castlereagh, had slit his own throat a decade earlier and Fitzroy would do the same in 1865. He was sufficiently worried about succumbing to melancholy in the desolation of Tierra del Fuego that he refitted the *Beagle* at his own expense with a belowdecks captain's cabin, removed as far as possible from Stokes's death chamber, and obtained permission to carry a paying guest who might fill the dual role of gentleman's companion and volunteer naturalist.[6]

Once the admiralty authorized a civilian passenger, the old-boy network of nineteenth-century British science did the rest. A science-minded

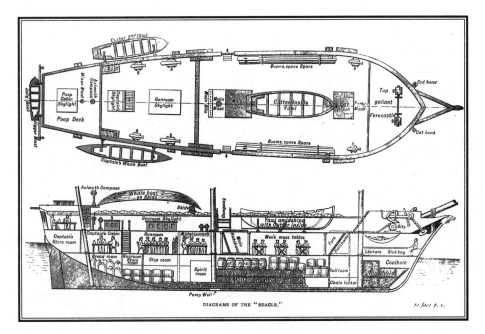

Diagram of HMS Beagle, *from Charles Darwin,* Journal of Researches *(1901 ed.)*
(Source: Used by permission of UGA.)

official in the admiralty contacted the eminent circle of scientists at Cambridge University, who recommended a recent graduate, Charles Darwin. He was not the first choice of the Cambridge circle; but the others all declined and the lot passed to young Darwin, who had nothing better to do anyway. An indifferent medical student at Edinburgh University, Darwin had moved to Cambridge for a proper liberal arts education in preparation for a career in the ministry at a time when the Anglican clergy served more of a social function than a religious one. At Cambridge he developed an abiding passion for natural history, regularly collecting specimens with the botanist John Henslow and once going on a summer field trip with the geologist Adam Sedgwick. The natural-theology bent of early nineteenth-century British science made such activities appropriate for a future gentleman cleric. An international collecting expedition would complete Darwin's training in natural history, his former teachers reasoned, just as it did for so many naturalists in that era before the formation of academic graduate schools. Although Darwin came from the landed gentry rather than the aristocracy, his social station and scientific training satisfied Fitzroy's peculiar requirements.

The invitation came first in a letter from Henslow, reaching Darwin at the family home in Shopshire, west-central England. He had recommended Darwin, he wrote, as naturalist-companion "to Capt Fitzroy employed by Government to survey the S. extremity of America—I have stated that I consider you to be the best qualified person I know of who is likely to undertake such a situation—I state this not on the supposition of your being a *finished* Naturalist, but as amply qualified for collecting, observing, & noting any thing worthy to be noted in Natural History." "Capt F.," Henslow went on, "wants a man (I understand) more as a companion than a mere collector & would not take any one however good a Naturalist who was not recommended to him likewise as a *gentleman*. I think you are the very man they are in search of."[7]

Darwin immediately wanted to go, but his father dismissed the voyage as yet another expensive dalliance. A trusted maternal uncle, the wealthy china manufacturer Josiah Wedgewood II, intervened by assuring the elder Darwin that "the pursuit of Natural History though certainly not professional, is very suitable to a Clergyman."[8] The father relented, as he usually did, and ultimately agreed to pay the full expenses for his son. Later, he also paid for his son's manservant, Syms Covington. "Gloria in excelsis is the most moderate beginning I can think of," Darwin soon exclaimed to Henslow. "What changes I have had: till one to day I was building castles in the air about hunting Foxes in Shopshire, now Lamas in S. America."[9] And not just llamas. He soon bubbled to a school chum, "It is such capital fun ordering things, to day I ordered a Rifle & 2 pair of pistols; for we shall have plenty of fighting with those d— Cannibals: It would be something to shoot the King of the Cannibals Islands."[10]

Darwin had one request of Henslow, Fitzroy and any admiralty official who would listen. Please let the ship return to England the long way around the globe: Darwin wanted a circumnavigation to his credit, not just a trip to South America. "They all think it most extremely probable, home by the Indian Archipelago: but till that is decided, I will not be so," Darwin explained in one letter.[11] Fitzroy promised to do what he could, and soon orders from the admiralty authorized the longer route home, which helped stretch the voyage to nearly five years. The final instructions to Fitzroy put the *Beagle* on course for the Galápagos Islands: "If he should reach Guaydaquil [Ecuador], or even Callao [Peru], it would be desirable he should run for the Galápagos, and, if the season permits, survey that knot of islands."[12] The season permitted.

The Galápagos did not feature prominently in planning for the trip. Darwin initially focused his attention on the mission's principal destination, southern South America, and that meant studying geology. "I was led to attend closely to several branches of natural history," he later said of the voyage, but "the investigation of the geology of all the places visited" came first.[13] This was particularly true of the arid regions of Patagonia and Chile, where naturalists had already surveyed the living animals and plants, but rocks and fossils lay bare for Darwin's examination.

In preparing Darwin for the expedition, Henslow had counseled his former student, "The Voyage is to last 2 yrs. & if you take plenty of Books with you, any thing you please may be done."[14] Sedgwick followed up by sending Darwin a reading list in geology and paleontology.[15] The list did not include *Principles of Geology*, the controversial new book by the London barrister-turned-naturalist Charles Lyell, who used legal reasoning and scientific evidence to revive and revise Joseph Hutton's evolutionary notions about geology. In Lyell's version, known as uniformitarianism, past geological processes acted much like present ones in gradually shaping the earth's features, organic life existed continually throughout a vast geological history and individual species appeared and disappeared over time. Sedgwick clearly intended the omission of Lyell's book: he had just used his presidential address to the Geological Society of London to denounce it. Following Georges Cuvier, Sedgwick maintained that periodic catastrophes (beyond anything presently known) shaped geological features and divided life into distinct, increasingly complex and separately created stages. The more open-minded Henslow corrected this deficiency in Darwin's reading list.

"When I started on the voyage of the *Beagle*, the sagacious Henslow, who, like all other geologists believed at that time in successive cataclysms, advised me to get and study the first volume of the *Principles*," Darwin recalled in his *Autobiography*, "but on no account to accept the views therein advocated." Henslow's reservation was forgotten once Darwin left the

British Isles and arrived at more geologically active ports of call. "I am proud to remember that the first place, namely St. Jago, in the Cape Verde Archipelago, which I geologised, convinced me of the infinite superiority of Lyell's views," he noted.[16] Here was a volcanic island of seeming recent origin that appeared to be the product of ongoing geologic forces. Fossils, fault lines and volcanos in South America confirmed Darwin's faith in Lyell, especially after he experienced a severe earthquake in Chile. "The world, the very emblem of all that is solid, has moved beneath our feet like a crust over a fluid," Darwin exclaimed in his *Journal*. "The most remarkable effect (or perhaps speaking more correctly, cause) of this earthquake was the permanent elevation of the land."[17] He had witnessed present geological forces sufficient to reshape the earth over time: these observations, coupled with Lyell's book, prepared Darwin to appreciate the spectacle of land being born in the Galápagos Islands; they also opened his mind to the parallel prospect of older biological species giving birth to newer ones. The second step, however, would take more time.

Though the connection between uniformitarian geology and organic evolution seems obvious in retrospect, in that both appeal to ongoing processes to explain nature, it did not immediately occur to many naturalists—certainly not to Lyell. His belief that the earth changed slowly led him to reject Cuvier's view of catastrophic extinctions and subsequent re-creations of life. Yet the alternative proposed by Cuvier's subordinate at the French Museum of Natural History, chevalier de Lamarck, of species evolving into other species in direct response to environmental changes, struck Lyell as equally repugnant. The fossil record did not display linking types, Lyell maintained, and he rejected as implausible Lamarck's notion that individual animals and plants could evolve new organs or traits through the use or disuse of existing ones. The newly published second volume of Lyell's *Principles of Geology*, which reached Darwin in South America, featured a detailed refutation of Lamarckian evolution.[18] Just as the first volume of the *Principles* encouraged Darwin to see geological evolution on the Galápagos Islands, the second volume discouraged him from seeing organic evolution there. For the time being, Darwin followed Lyell in accepting the occasional creation of species, their gradual spread to suitable locales and their eventual extinction due to changed environmental

Illustration of HMS Beagle, *from Charles Darwin,* Journal of Researches *(1901 ed.). Courtesy of University of Georgia Libraries*

Charles Darwin, shortly after he returned to England from the Galápagos, ca. 1839. Courtesy of G. P. Darwin on behalf of Darwin Heirlooms Trust. Source: English Heritage Photo

Chatham Mockingbird, cited by Darwin as evidence of evolution on the Galápagos in On the Origin of Species, *from Charles Darwin, ed.,* Zoology of the Beagle, 1838–43. *Courtesy of University of Georgia Libraries*

*U.S. Navy captain and Galápagos explorer
David Porter, from David Porter,* Journal of a
Cruise *(1822). Courtesy of United States Naval
Institute*

Porter's U.S. Frigate Essex, *1812. Courtesy of United States Naval Institute*

The Honorable Walter Rothschild in top hat riding a giant tortoise at his Tring Park estate. Courtesy of the Honorable Miriam Rothschild

C. M. Harris tending twenty-nine live Galápagos Tortoises from the Webster–Harris expedition destined for Walter Rothschild, 1898. Copyright © The Natural History Museum, London

Louis Agassiz (left) and Franz Steindachner (right) of the Hassler *expedition meet California Academy of Sciences President George Davidson on arriving in San Francisco from the Galápagos, 1872. Courtesy of Special Collections, California Academy of Sciences*

Elizabeth Cary Agassiz, founding president of Radcliffe College, around the time when she participated in the Hassler *expedition, 1872. Courtesy of Radcliffe Archives, Radcliffe Institute, Harvard University*

California Academy of Sciences director Barton Warren Evermann (left), who initiated the 1932 Templeton Crocker Expedition to the Galápagos, and Stanford University chancellor David Starr Jordan (right), who initiated the 1898–1899 Hopkins Stanford expedition to the Galápagos (both ichthyologists), at the San Francisco Fish Market, 1925. Courtesy of Alan E. Leviton, Special Collections, California Academy of Sciences

Entomologist F. X. Williams at the bow of the schooner Academy *shortly before its departure from San Francisco for the Galápagos, 1905. Courtesy of Special Collections, California Academy of Sciences*

R. H. Beck, leader of the 1905–1906 Academy *expedition, was also its chief photographer and bird collector. He is shown here photographing blue-footed boobies in the Galápagos. Courtesy of Special Collections, California Academy of Sciences*

The backbreaking work of hauling tortoises from the Galápagos highlands to the ship during the 1905–1906 Academy *expedition. Photo by R. H. Beck shows (from left to right) J. J. Parker, E. S. King, J. S. Hunter, F. X. Williams, W. H. Ochsner and A. Stewart. Courtesy of Special Collections, California Academy of Sciences*

250 giant tortoise specimens from the 1905–1906 Academy *expedition stored in the ruined California Academy of Sciences Museum after the San Francisco earthquake, 1906. Courtesy of Special Collections, California Academy of Sciences*

Ornithologist Harry Swarth during the Templeton Crocker Expedition to the Galápagos, 1932. Courtesy of Special Collections, California Academy of Sciences

Botanist John Thomas Howell during the Templeton Crocker Expedition to the Galápagos, 1932. Courtesy of Special Collections, California Academy of Sciences

American Museum of Natural History
President Henry Fairfield Osborn.
Photographer Julius Kirschner. Courtesy
of American Museum of Natural History

Galápagos explorers
Templeton Crocker (left)
and William Beebe (right)
aboard Crocker's yacht Zaca.
Courtesy of Special
Collections, California
Academy of Sciences

conditions. Initially, Darwin supposed that he saw this process, rather than organic evolution, in action on the Galápagos.

Getting to the archipelago took much longer than expected. The proposed two-year expedition had stretched to over three and a half years of meticulous surveying, and the party had yet to leave South America. Darwin tired of the adventure. Homesick, he now dreaded the circumnavigation he once craved. "But gracias a dios one month more & farewell for ever to Chili; in two months more farewell South America," Darwin wrote home in May 1835. "I have lately been reading about the South Sea—I begin to suspect, there will not be much to see."[19] By the time the *Beagle* finally left South America three months later, Darwin had built up some enthusiasm for the Galápagos Islands by reading Dampier, Colnett and Byron, but for little else except returning home. "I am very anxious for the Galápagos Islands,—I think both the Geology & Zoology cannot fail to be very interesting," he noted in a letter to his sister Caroline, but "nothing will be very well worth seeing, during the remainder of this voyage, excepting the last & glorious view of the shores of England." To Henslow, Darwin added, "In a few days time the Beagle will sail for the Galápagos Isds.—I look forward with joy and interest to this, both as being somewhat nearer to England, & for the sake of having a good look at an active Volcano."[20]

HMS Beagle *at sea, from Charles Darwin,* Journal of Researches *(1901 ed.) (Source: Used by permission of UGA.)*

The islands scarcely lived up to Darwin's expectations—certainly their volcanos disappointed him. Reflecting his preconceptions of the place, Darwin at first was more intrigued by its geology than by its biology. "The constitution of the whole is volcanic," he observed to open the Galápagos chapter of his published *Journal*. "There must be, in all the islands of the archipelago, at least two thousand craters. These are of two kinds; one, as in ordinary cases, consisting of scoriae and lava, the other of finely-stratified volcanic sandstone." Of the solidified outpourings from the former type of crater, he wrote with some enthusiasm, "Nothing can be imagined more rough and horrid than the surface of the more modern streams. These have been aptly compared to the sea petrified in its most boisterous moments." In marked contrast, he noted, the outpourings from the latter type of crater "have a form beautifully symmetrical," forming squat cones of solidified volcanic mud "without any lava." Fitzroy summed up the terrain more succinctly: "All the hills appear to have been the craters of volcanoes: some are of sandy mud, others are lava."[21] Yet when Darwin climbed these hills to look down into their craters, which he did whenever time permitted, he found them disappointingly quiet. "The craters are all entirely inert; consisting indeed of nothing more than a ring of cinders," he noted in his diary about Chatham Island. "When H.M.S. Blonde was here there was an active volcano in [Narborough] Island," but when the *Beagle* sailed between that island and Albemarle, "we saw [only] a small jet of steam issuing from a Crater."[22]

Darwin saw in the island geology precisely what Lyell's *Principles* had prepared him to see: land newly risen from beneath the ocean in the on-going cycle of uplift and erosion. The surface of the older, eastern islands had eroded enough for "some increase in the quantity of vegetation" since Dampier's visit, Darwin reported—enough even to sustain a self-sufficient settlement on Charles Island of more than two hundred persons, mostly political prisoners exiled from Ecuador then under the charge of English-born vice-governor Nicholas Lawson. The newer, western islands remained "sterile and incapable of supporting life," with both Albemarle and Narborough "covered with immense streams of black naked lava."[23] The variation from east to west caught Darwin's attention. It was, according to Darwin biographer Janet Browne, "the very stuff of Lyellian geology."[24]

As his expectation of erupting volcanos gave way to the reality of a bleak volcanic terrain, Darwin's fancy turned from the archipelago's geology to its zoology. Although he kept precise geological notes on each island visited, accounts of wondrous native animals soon crowded almost everything else out of his personal diary. This shift began on Darwin's second day in the Galápagos, when the *Beagle* moved from its desolate initial anchorage on northwest Chatham Island to the marine paradise of Stephen's Bay further down the coast. "The Bay swarmed with animals; Fish, Shark & Turtles were popping their heads up in all parts. Fishing lines were soon put overboard," Darwin wrote in his diary. "This sport makes all hands very merry." Greater wonders awaited on shore. "The black Lava rocks on the beach are frequented by large (2–3 ft) most disgusting, clumsy Lizards," marine iguanas. "Somebody calls them 'imps of darkness.' They well become the land they inhabit." Native birds drew his attention as well. "Little birds within 3 & 4 feet, quietly hopped about the Bushes & were not frightened by stones being thrown at them," Darwin observed. "I pushed off a branch with the end of my gun a large Hawk."[25]

The crew soon went in search of the islands' legendary tortoises, but found none that day. Scarcely three decades after whalers began frequenting the archipelago, excessive hunting had already taken a toll on the great beasts. Darwin reported that sailors and settlers looking for tortoises now had to scour the higher elevations. "It is said that formerly single vessels have taken away as many as seven hundred of the animals," he noted, but during the *Beagle*'s five-week visit, the crew managed to collect only four dozen. The sailors ate most of them, with only four small tortoises making it back to England as museum specimens.[26]

Darwin first encountered a tortoise four days after he arrived; his reaction betrayed his shift in focus from geology to biology. He had returned to examine craters at an earlier landing place on Chatham Island's northeast coast. "The day, on which I reached the little craters, was glowing hot, and the scrambling over the rough surface, and through the intricate thickets, was very fatiguing; but I was well repaid by the Cyclopian scene." The scene involved a zoological marvel rather than a geological one: "Two large tortoises, each of which must have weighed at least two hundred pounds."[27] Darwin added in his diary, "They were so heavy, I could

Island view, from James Colnett's report on Pacific whaling (1798)
(Source: Used by permission of UGA.)

scarcely lift them off the ground. Surrounded by black Lava, the leafless shrubs & large Cacti; they appeared most old-fashioned antediluvian animals; or rather inhabitants of some other planet."[28] From this point on, Darwin wrote more in his diary and *Journal* about tortoises than about anything else on the Galápagos, and only made passing references to island geology.

During his five weeks on the Galápagos, Darwin enthusiastically applied himself to collecting plant and animal specimens. For most of this time, however, the *Beagle* cruised among the islands with Fitzroy and the crew engaged in surveying the various coasts, shoals and offshore rocks—resulting in the most accurate maritime chart of the archipelago to that date. Darwin had but limited time to go ashore—only some brief visits to Chatham, Charles and Albemarle Islands and a week-long layover on James. All of his specimens came from these four major islands. Darwin made the most of his opportunities. "We slept on the sand-beach," he wrote of an overnight stop on Chatham Island with his servant, Covington, "collecting many new plants, birds, shells & insects." Throughout his longer stay on James Island with a party of five, "we all were busily employed during these days in collecting all sorts of Specimens."[29]

Darwin threw himself into this work. "I endeavored to make as nearly a perfect collection in every branch [of natural history] as time permitted," he later explained.[30] The young naturalist succeeded brilliantly. Assisted by Covington, Darwin skinned scores of birds and reptiles, collected numerous insects, preserved various fish and sea mammals and pressed several hundred plant specimens for identification by experts in England. Fitzroy and several others on board also collected specimens that they later donated to British scientific institutions. Darwin also had a chance to trek into the interior highlands on two islands—Charles and James—where low-hanging clouds produced what he generously described as "a tolerably luxuriant vegetation."[31] No previous collector had penetrated beyond the islands' arid coastal regions or even aimed at completeness there. These factors made Darwin's collections much more comprehensive than any before taken from the archipelago, especially for plants and land birds from the interior. Known endemic life on the Galápagos suddenly became more than tortoises, iguanas and boobies—though Darwin still paid considerable attention to these groups.

His Galápagos plant specimens exemplified the importance of all his collections to science. In a letter from the South Pacific, Darwin reported to his mentor Henslow about botanizing on the archipelago: "I worked hard. Amongst other things, I collected every plant, which I could see in flower, & as it was the flowering season I hope my collection may be of some interest to you." Following his return to England, Darwin's collection became the impetus and core for the first scientific listing and description of Galápagos plants, Joseph Hooker's 1847 "Enumeration," in which three-fourths of the species were attributed to Darwin. In their *Flora of the Galápagos Islands*, the definitive modern work on the subject, Ira Wiggins and Duncan Porter write that "Darwin's collection, the largest obtained up to that time, was the primary foundation for subsequent work on the vascular plants of the islands." Similar taxonomic significance attached to his collection of twenty-six species of land birds, twenty-five of which proved new to science and limited to the archipelago.[32] These specimens would later help point the way toward the theory of evolution, but at first even Darwin did not appreciate this.

To satisfy his immediate curiosity while still in the Galápagos, Darwin performed several crude "experiments" on the easily captured animals. He overturned tortoises to see if they could right themselves (they could) and rapped on their shells to see if this would draw them out (it did). Denouncing the former as "hideous looking" and the latter as "ugly," Darwin displayed particular delight in harassing the sea and land iguanas. Both types struck him as "stupid." He repeatedly tossed one sea iguana in water to prove that it favored land, and tried unsuccessfully to chase others into the sea. He pulled a land iguana by the tail simply to see its shocked reaction. "I opened the stomachs of several," Darwin wrote of both types, "and found them full of vegetable fibres"—leaves in the land iguana and seaweed in "their brothers the sea-kind." This created such a stinking mess in his cabin that his shipmates forbade it on board, but Darwin carried on ashore.[33]

No one could leave the passive birds alone: poking and prodding and throwing stones. In his field notes, Darwin attributed their "tameness" to the absence of cats and other predators. "I often tried, and very nearly succeeded, in catching these birds by their legs," Darwin wrote. Cowley had reported that Galápagos land birds knew no fear of people, and Darwin wanted to determine how much they had acquired over the intervening generations of human contact. Very little, he concluded, given the harassment that they suffered from sailors and settlers, but at least they now instinctively knew enough not to alight on people as they had once perched on Cowley and the other buccaneers.[34]

Darwin later interpreted the acquisition of an instinctive fear of humans as evidence for natural selection, but not yet. Selection theory would suggest that wild birds alighting on the buccaneers became easy dinner, while those with a native instinct to avoid a person's grasp survived to reproduce their kind. Only after those most easily caught became extinct did those hopping nearby become prey. Darwin described a Charles Island settler "sitting by a well with a switch in his hand, with which he killed the doves and finches as they came to drink. He had already procured a little heap of them for his dinner." In his *Journal*, Darwin used this episode to show how animals under selection pressure change gradually, suggesting that Galápagos birds soon would display an even greater instinctive fear of humans; that is where he left it in 1839. Only later would he reveal his growing conviction that this process, given enough time and acting on various characteristics, could evolve new species from old ones.[35] Darwin did not catch a glimpse of this new view while in the Galápagos, however. In his private diary, written on the spot in 1835, the story of the Charles Island settler illustrated the scarcity of water on the islands by showing how any available source (even one with a switch-bearing settler) would "draw together all the little birds in the country."[36]

With the second volume of *Principles of Geology* still fresh in his mind, Darwin followed Lyell in viewing species as variable only within prescribed limits. "There is a capacity in all species to accommodate themselves, to a certain extent, to a change of external circumstances," Lyell had written. "The entire variation from the original type, which any given kind of change can produce, may usually be effected in a brief period of time, after which no farther deviation can be obtained." Further adverse change in the environment simply killed off the species. "From the above consideration, it appears that species have a real existence in nature," Lyell concluded, "and that each was endowed, at the time of its creation, with the attributes and organization by which it is now distinguished."[37] Thus Darwin could conceive of Galápagos land birds quickly acquiring a limited fear of humans in response to the appearance of sailors and settlers, but readily accept Vice-Governor Lawson's prediction that humans would soon wipe out some of the archipelago's native species.[38]

Darwin conducted his investigation of Galápagos animals within the framework of Lyellian creationism. "Each species may have had its origin in a single pair," Lyell wrote in the *Principles*, "and species may have been created in succession at such times and in such places as to enable them to multiply and endure for an appointed period, and occupy an appointed space on the globe."[39] Fitting his work into this context, Darwin wrote in his diary about his visit on Charles, "I industriously collected all the animals, plants, insects & reptiles from this Island. It will be very interesting to find from future comparison to what district or 'centre of creation' the organized beings of this archipelago must be attached."[40] Most land mammals have limited ability to cross salt water, Lyell had noted, and so did not occupy oceanic islands remote from where they were created by nature or transported by humans. He specifically cited the Galápagos Islands as "examples of this fact."[41] Darwin confirmed the absence of land mammals on the archipelago, except for a small South American type of rodent, which he presumed had rafted there by chance, and an "English kind" of rat, probably off buccaneer ships, which he saw as "altered by the peculiar conditions of its new country."[42]

Lyell viewed reptiles as different in this respect from mammals. "By water," he wrote, "they may transport themselves to distant situations more easily."[43] So in his diary from the Galápagos, Darwin exalted, "These islands appear paradises for the whole family of Reptiles." More specifically, he noted in his *Journal*, "These islands are not so remarkable for the number of species of reptiles, as for that of individuals." Darwin suggested that some few reptiles floated to the archipelago from the nearest continental land mass (here speculating about the ability of reptile eggs to survive in sea water) and flourished there without mammalian competitors. "Of lizards there are four or five species," he noted; "two probably belong to the South American genus Leiocephalus." Other species of reptiles, such as the tortoises, might have originated in the Galápagos.[44]

Galápagos plants and land birds offered parallel tests of Lyell's theory on the creation and distribution of species. Wind or water could carry some of them to the islands; others simply first appeared there. "I shall be very curious to know whether the Flora belongs to America, or is peculiar," Darwin soon wrote to Henslow regarding Galápagos plant specimens. "I certainly recognize S. America in Ornithology," he added in a pocket notebook that he kept on the Galápagos; "would a botanist"? In a particular comment about a type of land bird that he found only on the central islands, Darwin wrote in his *Journal*, "It would appear as if this species had been created in the centre of the Archipelago, and thence had been dispersed only to a certain distance."[45]

The precise distribution of any species within the Galápagos did not then mean much to Darwin because he viewed all the islands as so similar that every fitting type would eventually spread over the whole group from its point of initial introduction or creation. "I therefore did not attempt to make a series of specimens from the separate islands," he later lamented. It did strike him as "singular" that distinct types of Galápagos mockingbirds occupied separate islands, and that even the archipelago's famed tortoises appeared to differ markedly from island to island, but at the same time he dismissed them as mere "varieties" of individual species, which neatly fit Lyell's views on such matters. Darwin left the Galápagos still a Lyellian creationist. He wrote in his diary about Albemarle, the largest of the islands, words that could have applied to any of them: "I should think it would be difficult to find in the intertropical latitudes a piece of land 75 miles long, so entirely useless to man or the larger animals."[46]

Darwin's thinking on the origin of species began evolving as he reviewed his Galápagos collections during his yearlong voyage back to England. He refocused his attention on the singular relationship among the mockingbirds from the archipelago. "I have specimens from four of the larger Islands," he jotted in the private ornithological notebook that he kept aboard ship. "The specimens from Chatham and Albemarle Isd. appear to be the same, but the other two are different. In each Isd. each kind is *exclusively* found; habits of all are indistinguishable." This observation carried great significance if those differing specimens represented distinct species. Darwin then inserted his recollection of a comment by Galápagos vice-governor Lawson "that from the form of the body, shape of scales and general size, the Spaniards can at once pronounce from which Isd. any tortoise may have been brought." Darwin seemed to ask himself, Why would God separately create on neighboring islands, or nature transport there, such nearly identical species? Could they instead have evolved on different islands from a common ancestor? "If there is the slightest foundation for these remarks," Darwin declared in these private notes made during his long voyage home, "the Zoology of Archipelagoes will be well worth examining; for such facts would undermine the stability of species."[47]

He pulled back from the precipice. An alternative explanation existed within the norms of early nineteenth-century science: although Lyell denied the evolution of different species, he granted wide variation in each species. So Darwin concluded, "When I see these Islands in sight of each other and possessed of but a scanty stock of animals, tenanted by these [mocking]birds but slightly differing in structure and filling the same place in Nature, I suspect they are only varieties."[48] These notebook jottings carried no date and Darwin never published them, but based on their context, historian Frank J. Sulloway places their writing between June 18 and July 19, 1836, after the *Beagle* had rounded the Cape of Good Hope heading for home.[49] By then it was too late for Darwin to go back to the Galápagos to test his ideas; he had to await expert examination of the specimens collected there without this hypothesis in mind. Thus when discussing the Galápagos mockingbirds in his 1839 *Journal*, Darwin added the universal lament, "It is the fate of every voyager, when he has just discovered what object in any place is more particularly worthy of his attention, to be hurried from it."[50]

Following his return to England in the fall of 1836, Darwin turned over the bulk of his natural history collections to experts for identification and classification, with the plants going to Henslow and most of the animals to the Zoological Society of London. Henslow tarried for years, preoccupied with teaching, other research and his growing family; but various gentlemen-scientists in the Zoological Society promptly went to work on its *Beagle* specimens. On January 4, 1837, scarcely a week after Darwin delivered his collection, several London newspapers reported a meeting of the society. "On the table," said the *Morning Harald*, "was part of an extensive collection of mammalia and birds, brought over by Mr. Darwin, who accompanied the Beagle in its late surveying expedition in the capacity of Naturalist, and at his own expense." The article noted that the eminent ornithologist John Gould "described 11 species of the birds brought by Mr. Darwin from the Gallapagos Islands, all of which were new forms."[51] The number grew on closer examination, so that the official record of the meeting told of Gould exhibiting "a series of *Ground Finches*, so peculiar in form that he was induced to regard them as constituting an entirely new group, containing 14 species, and appearing to be strictly confined to the Galápagos Islands." In particular, Gould "remarked that their principle peculiarity consisted in the bill presenting several distinct modifications of form."[52]

This revelation about Galápagos finches, many of which Darwin had not even identified as finches, revived his thoughts about species evolving through accumulated variations.[53] In effect, he asked himself, why else would so many similar species live so near each other? Such evolution, however, required isolation; otherwise the varieties should never develop into separate species, but more likely would merge through interbreeding. Darwin had labeled all these finches as collected on the Galápagos Islands, but not which island they came from. If finch species mingled freely, then they may have been created or introduced as distinct species. If they came from separate islands or were otherwise segregated, however, then they more logically evolved in isolation from a common ancestor.

Darwin tried to reconstruct the source for each finch species by consulting precisely labeled specimens collected by Fitzroy and others on the *Beagle*, but these could not confirm their geographic isolation. Indeed, Darwin later told the Zoological Society that he found the birds "indiscriminately associating in large flocks" and that all of them "subsist on seeds, deposited on the ground."[54] If true, both observations denied the isolation of separate species and the adaptation of their bills for specialized feeding. Darwin thus never used his finches as evidence for evolution in his published writings, though they clearly stimulated his thinking about it.

Those thoughts received greater inspiration when Gould ratified Darwin's earlier surmise that three of the Galápagos mockingbirds constituted distinct species. Darwin knew that these types were isolated from each other on separate islands and believed that they must have descended from common stock. If they now represented distinct species rather than mere varieties, then transmutation indeed had occurred. Gould communicated his findings to Darwin in March 1837.[55] At once, Darwin's ideas took flight.

"In July opened first notebook on 'Transmutation of Species,'" Darwin jotted in his diary sometime in 1837, identifying "species on Galápagos Archipelago" as the primary source "of all my views."[56] An accumulating series of personal notebooks and private essays followed, in which Darwin assembled his theory of evolution while hardly telling anyone about it. In these early arguments, the Galápagos Archipelago served as his idealized test case and its distinctive species as his best evidence. "My idea of Volc: islands. elevated. then peculiar plants created" by introduction and adaptation, Darwin hypothesized in his first notebook on the subject. "Yet new creation [of island species] affected by Halo of neighbouring continent."[57] In other words, plants and animals from the nearest landmass should colonize a newly formed island and thereafter evolve into new species fitting the island's environment and filling available niches. "So if islands formed near continent, let it be ever so different," Darwin wrote in a private 1842 sketch on evolution, "that continent would supply inhabitants, and new species (like the old) would be allied with that continent." Under creationism, in contrast, a good Creator should fashion new species to fit their locales without any alliance to ill-suited nearby types. Darwin emphasized this point in a private essay two years later: "Even the different islands of one such group are inhabited by species distinct, though intimately related to one another and to those of the nearest continent." This, he noted, "is so wonderfully the case with the different islands of the Galápagos Archipelago."[58]

Throughout the period from 1837 to 1844, Darwin looked to his Galápagos collections for evidence to support his theory. He welcomed the news that Gould ultimately identified all but one of his twenty-six different Galápagos land birds as new species confined to the archipelago. Equally exciting, Darwin learned early in this period that naturalists now considered the giant tortoise as native to the Galápagos and segregated into two distinct island-specific species—even though the *Beagle* specimens contributed nothing to this finding. He heard of it from the Zoological Society herpetologist Thomas Bell in 1837, but when a visiting French naturalist confirmed it to Darwin a year later, he exclaimed anew in a personal notebook, "Two new species of Tortoise come from Galápagos!!!" Previously, naturalists had viewed Galápagos Tortoises as varieties of a widely dispersed species, perhaps transported to the archipelago by aboriginal visitors, a view maintained by Fitzroy in his narrative of the *Beagle's* voyage. With evidence mounting for evolution on the Galápagos, Darwin pushed Henslow for conclusions about "the general character of the vegetation" there and asked the naturalist Leonard Jenyns, who had agreed to identify fish collected on the *Beagle* expedition, to look first at those from the archipelago.[59]

Darwin pulled together the available information on Galápagos species for his private 1844 essay. "Here almost every [land] bird, its one mammifer, its reptiles, land and sea shells, and even fish, are almost all peculiar and distinct species," he wrote, yet "they belong to the American type." Further, all of the archipelago's islands "are of absolutely similar composition, and exposed to the same climate; most of them are in sight of each other; and yet several of the islands are inhabited, each by peculiar species (or in some cases perhaps only varieties) of some of the genera characterizing the archipelago." He compared this with physically similar volcanic islands off Africa, also visited by the *Beagle*. "The Cape de Verde group, to which may be added the Canary Islands, are allied in their inhabitants (of which many are peculiar species) to the coast of Africa, in precisely the same manner as the Galápagos Archipelago is allied to America."[60]

Writing only to himself, Darwin then turned on his imagined opponents. "The creationist" must consider these "as so many ultimate facts," he observed. "He can only say, that it so pleased the Creator . . . that the inhabitants of the Galápagos Archipelago should be related to those of Chile, and that some of the species on the similarly constituted islands of this archipelago, those most closely related should be distinct: that all its inhabitants should be totally unlike those of the similarly volcanic and arid Cape de Verde and Canary Islands." This could be, Darwin replied to himself, "but it is absolutely opposed to every analogy, drawn from the laws imposed by the Creator on inorganic matter, that facts, when connected, should be considered as ultimate and not the direct consequences of more general laws."[61] In the privacy of his own mind and personal writings, Darwin had made his case for organic evolution.

The Galápagos Islands would carry Darwin only so far. After inspiring his vision that species evolve, they left him without clear guidance about how it happened. This question hounded Darwin for years. A generation earlier, Lamarck had proposed that individuals change in response to their environment and thus evolve new species over time, but the scientific community ridiculed him. Darwin initially supposed that, given an inherent tendency for individuals to vary, simply isolating a breeding population might generate a new species. "According to this view," he jotted in a mid-1837 notebook entry, "animals, on separate islands, ought to become different if kept long enough apart, with slightly differen[t] circumstances.—Now Galápagos Tortoises, Mocking birds; Falkland Fox—Chiloe fox.—Inglish and Irish Hare."[62] To see if this were so, he obsessively studied the process of variation among plants and animals in nature and under domestication. Evidence from the Galápagos could not help him here because he had visited the archipelago only for a few weeks—not nearly long enough to watch variation develop over time—and no other naturalist had investigated the matter there.

His inspiration came from reading Thomas Malthus's *An Essay on the Principle of Population* in the early autumn of 1838. Writing about people, not plants or other animals, Malthus asserted that humans have a natural tendency to reproduce at an unsustainably rapid rate. Competition among people checks the excess through famine, epidemic, war and the like, tending to leave the fittest to survive and reproduce their kind.[63] "Overwhelmed by the instant recognition that comes to a prepared mind, Darwin saw that what Malthus said about checks to fecundity in the human world rang emphatically true for animals and plants also," Darwin historian Janet Browne observes. "Darwin's moment of insights came when he caught at the idea that the ones who died would be the weakest and the ones who lived the strongest—or best adapted. Death, so to speak, could be a creative entity."[64]

Darwin would call the elimination of the weak *natural selection*: just as breeders preserve and propagate variation in domesticated plants and animals through selection, so did nature through a Malthusian struggle for survival. Over time, such selection could lead to new species and the wondrous diversity of life. "One may say there is a force like a hundred thousand wedges trying [to] force every kind of adapted structure into the gaps in the economy of Nature, or rather forming gaps by thrusting out weaker ones," Darwin wrote in a private notebook sometime late in September 1838, while he was still reading Malthus. "The final cause of all these wedgings, must be to sort out proper structure & adapt it to change."[65] Half horrified by its implications for traditional social values and religion, Darwin kept this theory largely to himself for more than a decade, working and reworking its every detail. "In a sense he was becoming his own William Paley," Browne remarks, "but a William Paley who reinterpreted all the myriad contrivances of the living world as an inevitable consequence of chance and change" rather than divine design.[66]

To the world, Darwin was becoming an increasingly prominent Victorian gentleman scientist. He oversaw the distribution of his *Beagle* specimens among British naturalists, which made him many friends, and produced a steady stream of scientific publications based on his voyage, including books on coral reefs (1842), volcanic islands (1844), South American geology (1846), barnacles (1851 and 1854), and the trip itself—his enormously popular *Journal of Researches* (1839 and 1845). During one notable week in 1839, Darwin gained election to the Royal Society of London and married his pious and proper first cousin, Emma Wedgwood (the wealthy Josiah's daughter). Despite his growing scientific status and financial security, he did not publicly announce his grand theory until 1858, when field naturalist Alfred Russel Wallace hit upon the same basic idea from his work in the Malay Peninsula. In the meantime, however, Darwin's publications increasingly set forth the Galápagos evidence that underlay his theory. In his treatise on volcanic islands, for example, Darwin discussed the archipelago's recent emergence from the ocean and its wholly volcanic character, concluding with the comment that "the islands are slowly clothed with a poor vegetation, and the scenery has a desolate and frightening aspect."[67]

Darwin slightly tipped his hand in the second edition of his *Journal*. The first edition (which he finished by mid-1837 and which closely follows his travel diary) largely reflected the author's earlier Lyellian views on the ori-

Finch beaks, from Charles Darwin, Journal of
Researches, *2d ed. (1845)*
(Source: Used by permission of UGA.)

gin and distribution of Galápagos species. He did insert a few sentences
(nowhere found in his diary) about Galápagos finches and the novelty of
plant and animal life on the archipelago, which he coyly called "a little
world within itself," but never betrayed the evolutionary conclusions that
he had begun drawing from these observations.[68] Yet in his mind, the novelty of Galápagos species pointed toward their evolution on the archipelago. Without evolution, they should remain like their American cousins.

Between the publication of the *Journal's* first edition in 1839 and its second in 1845, Darwin pressed for evidence supporting evolution from his
Galápagos plant specimens. He had labeled these specimens by island and
dearly wanted to know if they displayed his predicted evolutionary distribution pattern of novelty throughout, similarity to American types and
island-by-island segregation. After Henslow got nowhere with them in six
years, Darwin transferred the specimens to the more responsive Joseph
Hooker in 1843.[69] In the nick of time for inclusion in the 1845 *Journal*,
Hooker confirmed to Darwin, "The Florula of each Islet is 1/2 peculiar."
More critically, Hooker added, "The collection is *out & out* S. American,
& W. coast, but from the peculiarity of some genera & most species, I
should not have known where to put it, supposing Galápagos not to

xist."[70] Darwin was ecstatic.[71] Surely these plants must have come from the American mainland and evolved into peculiar species on the Galápagos. No other natural explanation was plausible.

For the *Journal*'s 1845 edition, the Galápagos chapter grew by half and became much more than a traveler's diary. The revised chapter featured Hooker's findings, complete with a new chart highlighting the number of plant specimens unique to each island and to the archipelago as a whole. Darwin also greatly expanded his discussion of Galápagos finches, adding a sketch of their differing bills and concluding with the blast, "Seeing this gradation and diversity of structure in one small, intimately related group of birds, one might really fancy that from an original paucity of birds in this archipelago, one species had been taken and modified for different ends." He noted similar novelty and diversity among Galápagos reptiles, insects, fish and shells, including a new reference to David Porter's 1813 observation of island-specific differences among Galápagos Tortoises. It is not simply the abundance of unique species on the islands, Darwin stressed, but "the circumstance, that several of the islands possess their own species of the tortoises, mocking-thrush, finches, and numerous plants . . . that strikes me with wonder." Here on this "little world within itself, or rather, a satellite attached to America, whence it has derived a few stray colonists . . . we seem to be brought somewhat near to the great fact—that mystery of mysteries—the first appearance of new beings on this earth."[72] To Hooker, in 1846, he added, "The Galápagos seems a perennial source of new things."[73]

In public, Darwin revealed nothing further of his thinking about evolution until shortly before publication of his *On the Origin of Species* in 1859. Yet he labored continuously on his argument during this period, assembling myriad examples and analogies that demonstrated a pattern of gradual organic evolution. None of these examples proved Darwin's theory in the sense of providing an observable instance of one species evolving into another; but cumulatively they showed that evolution more logically explained life than did special creation. As the examples multiplied, those from the Galápagos Islands lost their preeminence—but Darwin always credited them as his primary source of inspiration.[74] Of course, as the 1845 edition of his *Journal* conceded, God could have created unique species for the Galápagos Islands, but "why were they created on American types of organization?"[75]

Even as Darwin held back from further releasing these heretical views—once privately berating himself as "a Devil's chaplin" for his work—he sought confirming evidence.[76] He befriended Galápagos explorer Hugh Cuming (who then lived in London) and inquired about his collections; examined European herbaria and museum specimens from the archipelago; and urged later explorers to investigate the distribution of species there. He pleaded with the naturalist Thomas Edmonstone, who accompanied an ambitious 1845 British naval expedition to the Pacific, "Collect everything at the Galápagos, & attend particularly to the production of the *different* islands."[77] Unfortunately, Edmonstone died on that voyage and his Galápagos specimens became so mixed up that no one ever sorted out their places of origin.[78] More significantly for science, Darwin worked closely with Gould and Hooker in preparing their respective publications on Galápagos birds and plants, both of which identified the vast majority of Darwin's different specimens as representing newfound species endemic to the archipelago. Some naturalists disagreed, seeing them as mere varieties of widely distributed species, but Gould's volume countered that "the experience of all the best ornithologists must be given up" if the Galápagos mockingbirds were not separate species.[79]

Expeditions to the South Pacific continued to call on the Galápagos Islands during the mid-1800s, though with less urgency after the *Beagle* had done its work charting the archipelago and collecting its specimens. The original itinerary for HMS *Sulphur* (charged with continuing the South American coast survey wherever the *Beagle* left off) potentially included the Galápagos, but the ship ultimately sailed through the archipelago without stopping.[80] Several expeditions from France and one from Sweden tarried longer in the islands, but none of them left much impact on Galápagos science.[81]

Once naturalists had sorted through all of Darwin's collections from the Galápagos Islands, there seemed little else to do there except the mundane work of adding to the list of native species.[82] Nothing they had found struck them as useful enough to demand further study: no new agricultural products (such as breadfruit from elsewhere in the South Pacific) or rich supply of natural resources (except perhaps of guano, some Americans speculated).[83] Other places offered greater economic potential. Further, Galápagos biology and geology still did not fit British natural theology. Even Fitzroy, a staunch creationist to the end, puzzled in his *Narrative* about Galápagos Tortoises, "This animal appears to be well defended by nature; but, in truth, it is rather helpless, and easily injured."[84] Herman Melville, writing "The Encantadas" about this time, offered scant comfort for the natural theologian by describing these reptiles as purgatorial incarnations of wicked sea officers.[85] Until science saw in nature something beyond human utility or divine design, the archipelago would arouse little scientific interest. Then Darwin published the *Origin of Species* in 1859, and everything changed.

By the time Darwin finally published his epic book, he had so augmented his argument for evolution by natural selection that the Galápagos Islands assumed a secondary role in it. For example, he devoted twice as many pages to drawing analogies to natural selection from pigeon breeding as to presenting evidence for evolution from Galápagos species. Indeed, he mentions the archipelago on only six different occasions, all near the end, to illustrate three separate themes.

Each use of Galápagos evidence reinforced the book's central thesis. First, oceanic islands (like "the Galápagos Archipelago") contain a high proportion of species found nowhere else, many of them closely related to each other. "This fact might have been expected on my theory," Darwin wrote, "for, as already explained, species occasionally arriving after long intervals in a new and isolated district, and having to compete with new associates, will be eminently liable to modification, and will often produce groups of modified descendants." Second, isolated places sometimes lack an entire type of plant or animal, with its place filled by another. For instance, "in the Galápagos Islands reptiles . . . take the place of mammals." This too fit his theory, Darwin noted, because some types cannot reach such places and others evolve to fill the void. Third, the distinctive flora and fauna of islands often resemble those of the nearest mainland, even where physical conditions greatly differ. "Why should the species which are supposed to have been created in the Galápagos Archipelago, and nowhere else, bear so plain a stamp of affinity to those created in America?" Darwin asked. "I believe this grand fact can receive no sort of explanation on the ordinary view of independent creation; whereas on the view here maintained, it is obvious that the Galápagos Islands would be likely to receive colonists, whether by occasional means of transport or by formerly continuous land, from America."[86]

Darwin here used the Galápagos Islands to deal hammer blows to natural theology. God did not providentially create species ideally suited for the archipelago; instead, chance arrivals evolved to fit there. Nor did the islands easily fall within Paley's "happy world." They were places of fierce competition among individuals and species where only the fittest survive.[87] As a young Cambridge University student under the orthodox Henslow and Sedgwick, Darwin had admired Paley's *Natural Theology* more than any other assigned text. The book contributed greatly to "the education of my mind," he later explained, by showing him how to argue logically from premise to conclusion. "I did not at that time trouble myself with Paley's premises," Darwin added, but from his experience on the Galápagos Islands he found them woefully deficient.

"The old argument of design in nature, as given by Paley, which formerly seemed to me so conclusive, fails, now that the law of natural selection has been discovered," Darwin concluded. "There seems to be no more design in the variability of organic beings and in the action of natural selection, than in the course which the wind blows. Everything is the result of fixed laws."[88] Whereas the archipelago shed little light on Paley's world, it was a beacon in Darwin's. "When I visited during the voyage of H.M.S. *Beagle*, the Galápagos Archipelago," Darwin wrote, "I fancied myself brought near to the very act of creation. I often asked myself how these many peculiar animals and plants had been produced: the simplest answer seemed to be that the inhabitants of the several islands had descended from each other, undergoing modification in the course of their descent; and that all the inhabitants of the archipelago were descended from those of the nearest land, namely America."[89] Others would want to see this for themselves.

PART TWO

EVOLUTIONARY DEBATES

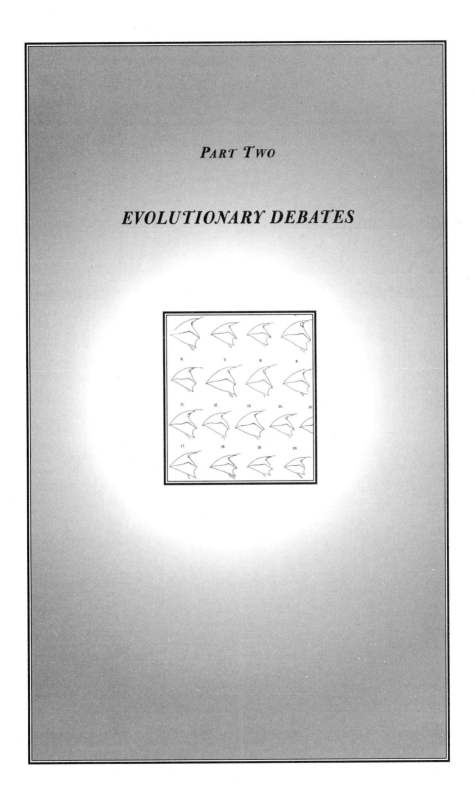

FOUR

COLLECTING ON
CLASSIC GROUND

ON THE ORIGIN OF SPECIES SHAPES the scientific view of life more than any other book. Darwin's work extended to biology the revolutionary concept of evolutionary origins that Lyell's *Principles* brought to geology, and did so with even greater eloquence. Like Lyell, Darwin could not prove his case by observing past processes, so he relied on analogy and hypothesis, drawing on evidence from the Galápagos Islands that had inspired his original conception.[1] By addressing the origins of life rather than land and the nature of humans rather than rocks, however, Darwin broached a much more sensitive subject.

Naturalists had long debated the history of life on earth. Linnaeus sought to trace the origins of each species to its first parents in Eden; Cuvier opted for successive acts of creation and extinction punctuating geologic history; Lamarck postulated evolution through progressive adaptation; and Lyell favored the occasional appearance and disappearance of species in response to changes in the local environment. Darwin offered yet another account of life's "splendid drama," as one participant in those debates termed it.[2] Unlike the theories of Linnaeus, Cuvier and Lyell, but like Lamarck's lowly regarded hypothesis, Darwin's alternative rejected the special creation of individual species. God might survive as the initial creator of Darwinian life, but if so, He acted remotely through ongoing laws of nature that scientists could investigate without ever looking over their shoulders for a divine designer.

Given that accepting Darwin's theory all but necessitated rejecting natural theology, scientists approached it contentiously. The ensuing debates inevitably uncovered gaps in Darwin's thinking, which necessarily engendered yet more controversy among scientists, many of whom appealed to nature for insight or evidence. Some of these turned back to the Galápagos Islands: What better place to look? Late in the nineteenth century, one prominent participant in these debates reflected that "the conditions of a critical experiment" to test how evolution works exist when "an organism A, with an environment or habitat A, is transferred to environment or habitat B, and after one or more generations exhibits variations B."[3] Another participant added, "The most favorable conditions are presented in the first place by small oceanic islands which owe their existence to volcanic outbursts."[4] The Galápagos offered just such conditions. The place that inspired Darwin could test and refine his theory.

On the Origin of Species raised two fundamental issues for science. First and foremost, it revived the scientific debate over whether organisms of one species evolved into those of another—a phenomenon then often called *transmutation*, to distinguish it from special creation. The publication of Darwin's book immediately conferred scientific credibility on the transmutation hypothesis and paved the way for its quick acceptance by most British and American biologists. Second, *Origin of Species* proposed a mechanism for transmutation, involving the natural selection of random, inborn variations—but this aspect of Darwinism encountered continued objections from scientists for more than a half century. Darwin himself waffled on mechanisms. He confided to one of his supporters as early as 1863, "Personally, of course, I care much about Natural Selection; but that seems to me utterly unimportant compared to the question of *Creation or Modification*."[5] In successive revised editions of *On the Origin of Species*, Darwin himself introduced so many Lamarckian mechanisms into the text that he all but obscured the Darwinian ones. As it turned out, evidence from the Galápagos Islands greatly supported the concept of transmutation but, given analytical tools available in the nineteenth century, revealed little about the mechanism involved; that would come later.

Despite Darwin's effective use of the geographical distribution of species on the Galápagos Islands in *Origin of Species*, the archipelago did not feature prominently in the initial debate over the book itself. Darwin's friend and scientific confidant, Joseph Hooker, did allude to his own study of Galápagos plants in reviewing *On the Origin of Species* for a leading British gardening magazine in 1859. "Such ideas as these," Hooker wrote about transmutation, "have occurred to many naturalists, when in endeavouring to classify their animals and plants, or to account for their distribution, they have found themselves forced to face the difficulty of accounting for their origin."[6] When Britain's leading comparative anatomist, Richard Owen, countered with a slashing review of the book a year later, however, he conspicuously avoided any reference to Darwin's telling arguments from the geographical distribution of species on the Galápagos. Instead, he highlighted the *Origin*'s professed reliance on biological evidence from "South America," carefully distinguished that from evidence derived from uninhabited islands (presumably the Galápagos), and then scorned it. "The grand argument from absence of mammalia & batrachians in Oceanic islands is probably felt to be strong by Owen as he has not ventured to impugn it," Lyell fairly gloated to Darwin.[7] Owen's review did concede that "the origin of species is the question of questions in Zoology." Despite his protests, the answer by scientists on both sides of the Atlantic was, increasingly, transmutation. "You will have the rare happiness to see your ideas triumphant during your lifetime," British naturalist T. H. Huxley exclaimed to Darwin in 1868.[8]

A minor publication by Huxley's closest friend in science, the British ornithologist Philip Lutley Sclater, illustrated the neat fit of Galápagos science into the new evolutionary regime. Before reading *Origins of Species*, as a gentleman scientist seeking to interpret the geological distribution of species, Sclater had divided the earth into six biologically distinct regions based on their bird life. He then spent a lifetime (when he was not organizing scientific associations, serving in high government posts or socializing with Britain's cultural elite) publishing over a thousand scientific papers that fit various species of birds into his overall scheme. As originally conceived, in 1857, Sclater's scheme conformed to the notion of biogeographic regions associated with the special creation of species in different places. Naturalists disagreed over whether this had happened in a handful of periodic explosions of life (as Cuvier maintained) or piecemeal throughout history (as Lyell countered), and whether the first members of a new species appeared in great numbers over its entire range (Cuvier) or in small numbers at either a center of creative activity or a few scattered sites (Lyell). By the mid-1800s, however, they generally rejected the traditional belief that all the species spread out from a single source, whether the Garden of Eden or Noah's ark.

The resulting scientific interest in the geographic distribution of species increased after Darwin gave the notion of biogeographic regions even greater meaning. Indeed, the codiscoverer of natural selection, Alfred Russel Wallace, built on Sclater's structure in his greatest book on evolution, the 1876 *Geographical Distribution of Animals*. Other naturalists did likewise. During the Victorian era, European scientists seemed nearly as intent on carving the earth into separate biological regions as their imperial governments were on carving it up politically and economically.[9] When Sclater got a chance to incorporate Galápagos bird life into his biogeography in 1870, he made the most of it.

As secretary of the Zoological Society of London from 1860 to 1903, a post he held concurrently with many others, Sclater oversaw publications for the institutional home of Darwinism in Britain and gained access to the flood of animal specimens flowing into its facilities, which were like a private club at the political and economic heart of the most far-flung empire in the history of the world. In 1869, these specimens included a selection of birds from Simon Habel of New York, the first collector to spend time at the Galápagos since publication of *Origin of Species*. "Dr. Habel stated that his whole collection embraced upwards of 300 specimens, referable to about 70 species, some of which he believed to be new to science," Sclater reported to the society.[10] He promptly reserved the entire collection for himself and his colleague Osbert Salvin to identify. They ultimately counted thirty-seven species (six new) among Habel's Galápagos specimens, duly assigned them to Sclater's South American biogeographic region and found that they fit an evolutionary pattern.[11]

Habel had stopped at seven islands during his time in the Galápagos, but collected birds on only three of them. "From Indefatigable Island," Salvin noted, "a large and important series of birds was obtained; and more

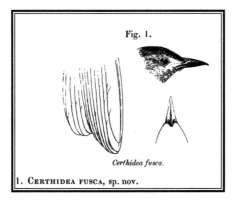

Fig. 1.

Certhidea fusca.

1. CERTHIDEA FUSCA, sp. nov.

Tree finch, from Sclater and Salvin's report (1870)
(Source: Used by permission of UGA.)

important still are [Habel's] collections from the more northern outlying islands of Bindloe and Abingdon. Neither of the last-named islands had before or since been visited by any naturalist." This piqued the interest of society naturalists because Darwin's theory of natural selection held that geographic isolation bred species. If Habel's specimens from these unexplored outer islands included new species, Salvin reasoned, they could help to confirm Darwin's views.[12]

Already drawn to walk in Darwin's footsteps less than a decade after publication of *On the Origin of Species*, Habel had gone to the Galápagos with the outer islands in mind. He traveled independently and with great difficulty, riding aboard Ecuadorian boats engaged in scavenging the islands' newfound but quickly depleted supplies of orchilla moss (used to make dye). The Zoological Society (which counted Darwin among its active members) shared Habel's enthusiasm. "The ground is classic ground," Salvin noted in his 1875 report to the society on Habel's collection, "and the natural products of the Galápagos Islands will ever be appealed to" by scientists studying the origins of new species.[13]

Salvin's final report on the Habel collection did little to resolve the issue of how evolution proceeded, however, because the species were not as neatly island-specific as Darwin's stress on isolation in evolution suggested they should be. Habel had found known birds (including the archipelago's puzzling finches) on previously unexplored islands and unknown birds on previously explored ones. "Mr. Darwin's views as to the exceedingly restricted range of many of the species must be considerably modified," Salvin concluded. Nevertheless, with respect to the occurrence of evolu-

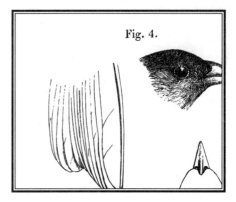

Woodpecker finch, from Sclater and
Salvin's report (1870)
(Source: Used by permission of UGA.)

tion (or "transmutation"), the report spoke with confidence: "The Birds that are now found [on the Galápagos], being related to American birds, must have emigrated thence and become modified by the different circumstances with which they became surrounded. The oldest immigrants seem to be indicated by their generic difference from their continental allies, the more modern comers by their merely specific distinctness, and the most recent by their identity with birds now found on the adjoining continent." Sclater's biogeography thus gained a historical dimension. Never mind that he could not explain how "different circumstances" caused transmutation or why similar species coexisted—Salvin saw evolution in action among Galápagos birds. By 1875, just sixteen years after Darwin's theory was published, he could not see them any other way.[14]

A further Darwinian influence marked both this report and one on a British navy expedition that stopped briefly at Charles Island in 1870: the destruction of native species by those introduced. "With the tortoises the mischief has all but been accomplished," Salvin's report observed. "Pigs now roam in their haunts, destined to destroy their eggs and young whenever and wherever they can find them." Similarly, the 1870 navy expedition found that tortoises "have almost disappeared from Charles Island," replaced by "several thousand head of wild cattle, besides pigs and goats." The feral animals also destroyed vegetation and habitat for native birds, endangering their survival as well.[15]

In *Origin of Species*, Darwin found this same process at work on the Falklands and other oceanic islands following the arrival of humans, and used it both to illustrate natural selection and to debunk the creationist

view of God designing species to fit their locales. The triumph of introduced species over native ones exemplified the survival of the fittest, he
reasoned, and belied any notion of a benign creation. Humans, of course,
participated in this process. For example, the navy report noted that orchilla scavengers had repopulated the old Charles Island settlement,
putting added pressure on native species. Further, sailors with the navy expedition eagerly hunted the naive birds for food and took potshots at anything else that caught their fancy. "We were rather startled on our way
back to the ship by the appearance of a shoal of very large porpoises all
around the boat, which gave good sport to a rifle," the navy report matter-
of-factly stated about the expedition's bloody departure from the archipelago. "Several were hit, but time would not permit us to stop and make an
addition to our bag."[16] Neither report made the Galápagos sound much
like the "happy world" of Paley's natural theology. It was more a part of
Tennyson's "nature red in tooth and claw." Its native animals would acquire
a healthy fear of humans or be replaced by introduced ones with that trait,
observers have predicted ever since.[17]

In a rearguard action, Darwin's leading opponent among nineteenth-
century naturalists, Louis Agassiz, made one final attempt to wrest the
Galápagos from the evolutionists. Brilliant, energetic and ambitious,
Agassiz studied with Cuvier in Paris during the 1830s and gained international fame among scientists by deducing from glacier formations in his
native Switzerland the concept of past ice ages. He applied this concept to
account for the massive extinctions found by Cuvier to punctuate the fossil record and to best Darwin in explaining the formation of geological
moraines.

Moving to the United States in 1846, the charismatic Agassiz—a gifted
huckster in a new nation on the make—quickly came to dominate America's small but growing scientific community from his twin posts as zoology professor at Harvard and founding director of the university's
Museum of Comparative Zoology. There in 1856 (three years before the
appearance of *Origin of Species* made it obsolete) Agassiz published his
long-awaited *Essay on Classification*, which elaborated his Cuvierian view
of multiple creations. Individual organisms exhibited too much complex-
ity in themselves, their relationship to each other and their fit into the environment, Agassiz reasoned, to have ever existed other than as they now

live. "It would follow that all animals and plants have occupied from the beginning those natural boundaries within which they stand to one another in such harmonious relations," his *Essay* proclaimed. Thus the species of each geologic epoch must have originated at once and in great numbers. "Pines have originated in forests, heathers in heaths, grasses in prairies, bees in hives, herring in schools, buffaloes in herds, men in nations," he wrote.[18] This concept left no place for either Noah's ark or organic evolution: for Agassiz, life had sprung forth in toto upon the thaw of each ice age and continued without significant change until the onset of the next killer freeze.

Obstinate to a fault, Agassiz clung to his theory of multiple creations until his death in 1873, fighting Darwinism at every turn. His battle was waged against materialism in science rather than for any particular religious creed. "He never identified with a sectarian religious persuasion," Agassiz's biographer Edward Lurie notes, "but was more fundamentally devoted to an idealistic romanticism that saw the power of the Creator exemplified in all flora and fauna."[19] According to Agassiz, "Beings do not exist in consequence of the continued agency of physical causes, but have made their successive appearance upon earth by the immediate intervention of the Creator." In classifying organisms into species and higher groups, naturalists "only translate into human language the Divine thoughts expressed in nature in living things." Agassiz stressed that such manifested thoughts—found in the structure, relationships and geographical distribution of organisms—reflected design: "They show the omnipresence of the Creator."[20] As the scientific world turned against him on these matters, Agassiz fought all the harder to assert his position—his struggle culminating in two grand expeditions to the very source of Darwin's evidence: South America and the Galápagos Islands.

The first expedition, conducted in 1865, got no further than Brazil, where Agassiz and twelve Harvard assistants (including the future philosopher William James) collected fish in the Amazon River. "I am often asked what is my chief aim in this expedition to South America?" Agassiz stated at the time. "The conviction which draws me irresistibly, is that the combination of animals on this continent, where the faunae are so characteristic and so distinct from all others, will give me the means of showing that the transmutation theory is wholly without foundation in facts."[21]

Amazon fishes attracted Agassiz's attention because he assumed that, under the doctrine of evolution, the same physical environment (or material causes) must produce the same living things. Finding different species of fish in otherwise similar pools of the Amazon, he reasoned, would undermine the theory of evolution. He compared these pools to islands of an archipelago, and found more different types of fish in them than even he dreamed possible. Scientists had identified only 100 species of fish in all of South America before Agassiz, yet he found over 2,200 species of them in Brazil alone. "Thus Agassiz was not only finding the very same 'great facts' Darwin discussed in the *Origin*," historian Mary Winsor observes, "he was seeing an even more clear and fine-grained instance of them, as if instead of the dozens of species on the Galápagos one had to do with an archipelago of islands stretching across hundreds of miles and populated by thousands of species of finches."[22] Yet Agassiz saw his findings as undercutting the theory of evolution. If, as he supposed, seemingly identical pools should produce seemingly identical species under the natural laws of evolution, the fantastic abundance he found could only arise because the Creator delighted in diversity. Darwinists would reinterpret his findings to support their view that geographical isolation in separate pools (or islands) facilitated the development of different species.

Agassiz tried again six years later, with an even more ambitious expedition that carried him by steamer around South America to the Galápagos Islands and brought him home by way of San Francisco and the new transcontinental railroad. On his 1865 expedition to Brazil, Agassiz traveled by commercial vessels, with the trip funded through private contributions ostensibly aimed at building the collection of the Museum of Comparative Zoology (which in fact the trip did, rather overwhelmingly). For the 1871 voyage, Agassiz had at his command the newest vessel in the U.S. Coast Survey fleet, the iron steamship *Hassler*, equipped with experimental deep-sea dredging equipment and provided to him at no cost by the head of the Coast Survey, Benjamin Pierce. The ship needed to go to California anyway, and Pierce hoped that Agassiz could make good use of the transport. "Would you go in her, and do deep-sea dredging all the way around?" Pierce asked Agassiz.[23] The sixty-four-year-old naturalist, still on the mend from a serious stroke, grasped at the chance to gather new evidence for his battle against Darwinism, even if it meant sailing through some of the most treacherous waters on earth. Darwin expressed respect for his adversary's determination. "What a wonderful man he is to think of going through the Straits of Magellan," Darwin wrote.[24]

Agassiz outlined his motive for the expedition in a letter to Pierce and in shipboard lectures to his assistants. "If this world of ours is the work of intelligence and not merely the product of force and matter," Agassiz wrote to Pierce before the voyage, "then we may expect from the greater depth of the ocean representatives resembling those types of animals which were prominent in earlier geological periods." Agassiz reasoned that if he could dredge living fossils from the deep sea—where they might have survived a cataclysmic ice age—then it would show that organisms did not evolve over time. Further, if he could correctly predict what basic types of as-yet-unknown organisms should live on the ocean floor, then an intelligible design must therefore govern life—and so Agassiz supplied Pierce with a detailed list of his expected haul. Finally, he took along copies of Darwin's writings to look for discrepancies between what he saw and what Darwin reported.[25]

Continuing a pattern of collaboration begun on the Brazil voyage, the naturalist's accomplished wife, Elizabeth Cary Agassiz, accompanied the expedition and joined in pummeling Darwin in print. As both a regular contributor to the *Atlantic Monthly* magazine and a pioneer of women's education at Harvard (she would later become the founding president of Radcliffe College), Elizabeth had her own audience for her gracefully worded brickbats.

As it turned out, the *Hassler*'s new dredging equipment did not work well enough to raise any evidence for or against the Agassizes' polemics, but the couple made the most of their ten-day stopover on the Galápagos. From first sight, they saw it as a place of benign creation.

"On a lovely day in June, 1872, we were approaching Charles Island in the Galápagos group," Elizabeth Agassiz reported in the *Atlantic Monthly*. "A marvelous school of porpoises, to be counted by hundreds, or perhaps by thousands, formed our escort. . . . Crowding about the bow of the ship, springing and jumping yards at a time, tumbling over one another, turning somersaults, they seemed to be having a great jubilee. One must be very familiar with the ocean to recognize the fact that there is as gay, as tumultuous, as enjoyable a life for animals in the sea as on land." Even among the Galápagos Islands, the Agassizes saw the "happy world" of Paley's natural theology. Elizabeth went on to relate "the mere delight in living" of whales, fishes, sea anemones, starfish—even corals. "They have an excellent time in their way," she affirmed.[26] Such observations resounded as an indictment of Darwin's Malthusian world where only the fittest survive.[27]

The delights of the Galápagos did not end at the coastline. The Agassizes landed on five different islands to examine geological formations and collect specimens for the Museum of Comparative Zoology. "Here we entered upon a truly wonderful lava region," Elizabeth Agassiz exalted. Louis "enjoyed extremely his cruise among these islands of such rare geological and zoological interest." The entire party shared his joy in the place. "I made all the artistic work, skinned all the birds and seals taken, collected shells etc," one museum assistant recalled, "but it was a continual pleasure." For her part, Elizabeth Agassiz especially enjoyed seeing flamingos one day. "Their attitude in the water is full of ease and grace," she wrote. "Shall I confess that beautiful as they were, and seemingly unfit for common uses, we dined on roasted flamingo that evening. Very tender and delicate it was, and of a delicious flavor."[28]

In a letter to Pierce and a report of the expedition in the British science journal *Nature*, Louis Agassiz used observations from the Galápagos Islands to directly challenge Darwin and his ideas. Some of these jabs assumed familiarity with Darwin's work. In *Journal of Researches*, for example, Darwin predicted that Galápagos animals would, by natural selection, acquire a fear of humans. "The tameness of these seals and of many of the land birds was very surprising," the *Nature* article reported. "I repeatedly put my fingers within a half an inch of little yellowbirds and phoebes," or about as close as Darwin had gotten to them. Darwin also had described Galápagos marine iguanas as "hideous-looking creatures" and suggested that they had evolved from land iguanas because they instinctively sought land, rather than water, when frightened. "I was prepared for something hideous, and was agreeably disappointed," Agassiz retorted. "In another respect our experience differed from Darwin's, for we sometimes had no difficulty in frightening them into the water."[29] So much, the report implied, for this indicator of evolution.

These passing slights simply set the stage for Agassiz's main argument against evolution on the Galápagos. "It is most impressive to see an extensive archipelago, of *most recent origin*, inhabited by creatures so different from any known in other parts of the world," he wrote to Pierce. Indeed, he stressed, some of these islands appear positively new. "Wherefrom, then do their inhabitants (animal as well as plants) come?" Agassiz asked. "If descended from some other types, belonging to any neighboring land, then it does not require such unspeakably long periods for the transmutation of species as the modern advocates of transmutation claim." After making the same point, the report in *Nature* concluded, "Darwin's hypothesis of gradual variation of species, and the natural selection for preservation of those whose variation were favorable to them in the struggle for life, seems to me to have few facts to sustain it, and very many to oppose it."[30]

Agassiz hardly needed to go to the Galápagos Islands to level these charges against Darwinism. By 1872, most evolutionists—even Darwin, grudgingly—acknowledged that natural selection alone could not evolve species fast enough to account for existing life, especially because physicists then placed the earth's age at only about 100 million years.[31] Yet the archipelago offered Agassiz a polemically effective pulpit for highlighting the issue. "Agassiz, of course, could have no knowledge of the power of mutations to effect rapid change, or that dominant traits could change populations in a relatively short period of time," his biographer writes. "But Darwin did not know of these evolutionary mechanisms either, and given the outmoded genetics he had advocated, the Agassiz position was a perfectly plausible one."[32] The topic required more study, Agassiz wrote to Pierce, and the Galápagos Islands offered an ideal location for that research. If evolution happened so rapidly there, Elizabeth Agassiz added in the *Atlantic Monthly*, "Then the transition types should not elude the patient student." Louis Agassiz issued a challenge aboard the *Hassler* as it steamed from the Galápagos: "The archipelago offers at present a fine opportunity for a naturalist, who desires to make a residence here for several years, and thoroughly explore their structure and their productions, to throw a strong light upon the great modern question of the origin of species."[33]

Agassiz died in 1873, a year after his return from the Galápagos Islands, the last great American naturalist to defend the special creation of species. By that time, the noted American paleontologist Edward Drinker Cope could describe transmutation as an "ascertained fact" even though he personally favored Lamarckian mechanisms over Darwinian ones.[34] Evolution dominated scientific thought in Britain and northern Europe as well. "Indeed," historian Peter Bowler concludes, "so great was Darwin's success in promoting the general idea of evolution that it is doubtful if more than a handful of biologists still accepted creationism as a viable option after the 1860s."[35]

Yet these first-generation evolutionists did not uniformly toe the Darwinian line. An explicitly anti-Darwinian school of evolution gained a foothold within British science during the late nineteenth century and clearly dominated biological thought in the United States during the period.[36] Evolutionists on all sides looked to the Galápagos for support.

Agassiz's legacy played a central role in this debate. He had trained a generation of America's leading naturalists, and even though most of them came to accept evolution, they often did so with a distinct bias against Darwinian mechanisms. At a technical level, Agassiz had raised questions about how evolution could operate with sufficient speed and direction to generate the diversity of life found on the Galápagos Islands and elsewhere. Some of his students sought the answers in neo-Lamarckian mechanisms that allowed organisms to adapt to changes in the environment, and pass those acquired characteristics on to their descendants. Others favored a role for either an external God or internal vital forces in directing orderly organic development: the former position became known as theistic evolution and the latter as orthogenesis. Thus there were at least three scientific alternatives to Darwinism among evolutionists.

At a philosophical level, Agassiz challenged the very notion that life could result from purely material forces. He saw design and purpose in nature, and so did many of his students. Neo-Lamarckism, theistic evolution and orthogenesis could infuse evolution with design, or at least purpose. "Rather than see their former mentor as a barrier to the spread of evolution in America," historian Ronald L. Numbers observes about Agassiz, "some of his former students portrayed him as a Darwinian John the Baptist, who 'prepared the way for the theory of evolution.'" Bowler adds that "Agassiz's disciples provided the kind of organizational network" that spread non-Darwinian theories of evolution throughout American science by the 1870s—a development soon mirrored in Germany and, to a lesser extent, Britain.[37] Many of these American scientists never forgot their teacher's admonition that the Galápagos Islands held clues to the puzzle of evolution. One of the greatest of them, David Starr Jordan, eventually rose to such eminence, as founding president of Stanford University and seven-term president of the California Academy of Sciences, that he could implement Agassiz's call to send naturalists for extended stays on the Galápagos. That, however, would take time. For the moment, focused collecting trips had to suffice.

Lofty debates over evolution aside, these late nineteenth-century collecting trips to the Galápagos Islands were driven by a more mundane purpose: gathering specimens for natural history museums. In an era before research universities and graduate education, Agassiz had secured private donations and government funds for advanced study in biology at Harvard through the Museum of Comparative Zoology. Its popular exhibits attracted public and private support, which also financed research. Even grander natural history museums arose elsewhere in the United States during the decades of prosperity following the Civil War: the American Museum of Natural History in New York, the Smithsonian Institution's National Museum in Washington and the California Academy of Sciences museum in San Francisco. Modern natural history museums also flourished in major European capitals, where imperial governments displayed biological trophies from their expanding colonial empires.

Britain laid scientific claim to the Galápagos as part of the Pacific, where it ruled so many islands; for the United States, the archipelago was a South American dependency subject to a sort of biological Monroe Doctrine. Other scientific powers with lesser claims in the region, such as France and Germany, occasionally intruded as well. The resulting competition yielded some of the best science of the day. Just as the great public museums became the centers for discovery and display of astounding prehistoric fossils, they also took the lead in the identification and classification of living species, and in studying their geographical distribution. All these topics raised issues about evolution.[38]

Fossilized dinosaurs and other ancient giants stood as centerpieces of many natural history museums. These specimens spectacularly displayed the effects of evolution or the manifestations of creation, depending on the viewer's perspective. In either case, they attracted the attention of museum curators and audiences alike. As museum collectors sought and found more fossilized remains, they also looked for "living fossils"—species from earlier geologic epochs that somehow survived in dark jungles, lofty mountains, oceanic islands or deep seas. British fiction writer Arthur Conan Doyle wrote a popular novel about such a "lost world," where dinosaurs persisted. Doyle set his lost world in South America, but soon after some naturalists found theirs offshore on the Galápagos Islands.

Few living fossils attracted more attention than the giant land tortoises found by European explorers only on the Galápagos Archipelago and on such Indian Ocean islands as the Mascarenes. Everyone had known, of course, that these giants haunted the Galápagos ever since the Spanish first found the place and named it after them. This belief changed as paleontologists began uncovering fossilized remains of giant tortoises throughout the world and zoologists detected significant differences between Galapàgos and Indian Ocean types, and lesser ones among those from each place. By the mid–1800s, many natural history museums possessed giant tortoise remains, but paid them little attention.

"The causes of the indifference with which these remains were treated are twofold," British Museum Keeper of Zoology Albert Günther observed. "First, the all-absorbing interest centered in the bird-remains; and, secondly, the belief that the bones were those of a [single] still-existing gigantic species of Tortoise" carried by sailors to different islands. This belief changed as paleontologists began uncovering fossilized remains of giant tortoises throughout the Northern Hemisphere and zoologists detected specific differences between Galápagos and Indian Ocean types, as well as lesser differences in both places. Günther put together the entire puzzle in a four-part monograph presented to the Royal Society of London in 1874. He unwittingly loosed a giant-tortoise craze among museum curators, who suddenly saw the beasts as living fossils from the early Tertiary Period surviving on a few remote islands where mammals could not reach them. Introduced competitors and hungry sailors had all but eliminated them even in those out-of-the-way places, Günther lamented, except on the least visited islands in the Galápagos group and one particularly rugged Indian Ocean atoll named Aldabra.[39]

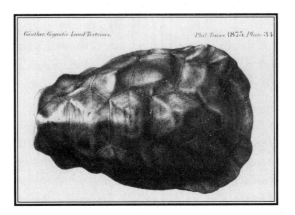

Saddleback tortoise shell, from Albert Günther's
treatise (1875)
(Source: Used by permission of UGA.)

Günther examined tortoises and their remains from various European museums and zoos in his initial research, he explained, "but the majority of specimens are young, or fragmentary, or without any history; and there will be found scarcely one with an indication of the particular island from which it came!" Identifying the place of collection had carried little importance so long as collectors thought that all tortoises came from a single stock scattered by sailors. Yet Günther's "closer examination and comparison" revealed "a multiplicity of species," with each species coming from a different island. "The results of these researches were startling, and may arrest the attention of the zoologist all the more, as the facts elucidated bring us face to face with the mystery of the birth and life of an animal type."[40]

"If a complete set of examples from every island had been secured for examination," Günther ventured, then the origins of these species could be determined.[41] Lacking such evidence, he held back from opining on that fundamental issue. Still, he did the best he could with the specimens available to him. In his monograph, Günther described in minute detail four separate species from the Galápagos specimens and assigned them to different islands. He also described separate but similar species from recent remains gathered on various Indian Ocean islands and from fossils collected throughout the Northern Hemisphere.

Such caution over origins fit Günther's scientific style. He worked for Darwin's British nemesis, British Museum superintendent Richard Owen, and throughout his long and distinguished career maintained a studied indifference to the theory of evolution. Like Agassiz, Günther in fact viewed the persistence of living fossils, such as Galápagos Tortoises, as evidence for creationism.[42] Yet his work suggested to other naturalists that these giant beasts might hold the key to evolution and bought them a prominent place in museum displays of prehistoric animals.[43] It also sent naturalists and museum collectors to the Galápagos in search of surviving tortoises, and gave a reason for haste. "The causes of their extinction having been at work for so long a time," Günther warned in his 1874 monograph, "that "what happened in the Mascarenes has commenced in the Galápagos."[44]

His pleas produced results. Less than a year after Günther read his monograph to the Royal Society, the British Admiralty ordered its next ship in the region, HMS *Peterel*, to stop in the Galápagos for tortoises. "These animals are extinct in Charles Island; and only a very few individuals are supposed to survive on Chatham Island. In Hood, James, and Indefatigable Islands the numbers are so reduced that they are no longer hunted," Commander W. E. Cookson of the *Peterel* reported on his return. "Albemarle and Abingdon are the only remaining islands in which they have ever been found." So he sent his sailors ashore on those two islands. They gathered thirty tortoises on the big island of Albemarle and brought back reports of many more holding out in its remote highlands. "In Abingdon Island," the commander warned, however, "I believe they are doomed to destruction directly [from] the orchilla-pickers." Accordingly, he ordered his sailors to collect every tortoise they could find there. Except for some eaten on the return voyage (which made "excellent soup"), Cookson turned over the entire catch, together with assorted other zoological specimens, to Günther at the British Museum.[45]

Günther expressed his appreciation and promptly published scientific descriptions of these additions to the museum's collection. "But for Commander Cookson's timely visit, the Abingdon Tortoise would, in all probability, have disappeared, unknown as if it had never existed," Günther wrote.[46] Apparently he believed that a specimen preserved at the British Museum offered immortality to the species. He certainly exhibited greater concern for expanding museum collections than for preserving species in the wild. Many naturalists felt this way, and they could command resources for their collecting missions. With rising public and private support for natural history museums, it became a happy world after all, at least for naturalists.

In the wake of the *Peterel*, a series of passing navy ships called on the Galápagos Islands over the following years to collect tortoises and other museum specimens. For example, the British Admiralty ordered the great ironclad HMS *Triumph* to stop there in 1880, though pressing military duties permitted it only one landfall. Unfortunately, the ship's captain chose the frequently visited Charles Island for that landing, apparently because he knew that Darwin had found tortoises there. Darwin, of course, had warned that human settlement on the island threatened to wipe out these tortoises, and the *Triumph*'s captain found that Darwin's grim prediction had come to pass. "During our stay on the island," he reported, "the only signs of a tortoise that was observed was the shell of a dead one in the vicinity of the settlement." The settlers too had gone, leaving their livestock to run wild. "There stood the huts," the captain observed, "silent as the tomb, a silence that was only disturbed by the birds which fluttered around us fearless and confiding. So tame were these, that they even alighted on our gun-barrels"; and so, with "praiseworthy zeal and energy," his dutiful officers collected birds.[47]

The resulting specimens went to the Zoological Society of London—a logical repository, given the ornithological expertise of Sclater and Salvin. So many trained collectors had already covered this particular ground, however, that the *Triumph*'s zealous amateurs netted no new species. Although any strong sailor landing on the right island could gather tortoises, by this time it took discretion to pick out strange birds in familiar places. Salvin used his report on the *Triumph* collection to point future Galápagos researchers toward less-traveled islands. "Much remains to be done," he stressed. "Not only are the important islands of Hood, Tower, and Albemarle almost untouched, but outlying rocks such as Wenman and Culpepper ought not to be wholly neglected."[48]

Salvin could not overlook the *Triumph*'s failure to bring back tortoises. "What is required of future explorers is the search in each island not only for the few lingering individuals that may still survive, but for any remains," he advised. "No time is to be lost, for, after the extinction of these animals which must ensue in a short time, all traces of them must follow at no distant period." These and other Galápagos specimens still held the key to evolution. "The Galápagos are strictly 'Oceanic Islands,' and owe their fauna and flora to the gradual immigration of species from Continental America," Salvin explained. "With the lapse of time, and the altered conditions amid which surviving species have found themselves, modifications in their form and color have been accumulated at a comparatively rapid rate." Researchers and museums, he asserted, needed specimens from the islands to investigate and illustrate this process.[49]

Collectors from other European countries followed the British to the place, also in search of museum specimens. An Italian navy ship stopped at tiny Duncan Island in 1884, where its sailors stumbled upon a trove of living tortoises. A French navy vessel anchored at Charles Island three years later, but its officers and crew only revisited the abandoned settlement. "One sees a few lizards," the French captain complained, but as for tortoises, "I, myself, have not been able to see a single specimen." The 1880s also included two visits to the archipelago by German naturalist Theodor Wolf, then engaged in a geologic survey of Ecuador, but his collections were lost in storage. Wolf did publish his observations of the islands, depicting them as the exposed tops of oceanic volcanos with a distinctly different composition from the volcanic mountains of South America. The older, larger islands had three distinct belts or levels, he noted: rocky, barren coastal regions; a higher vegetative zone where moist trade winds created fertile soil from volcanic rock; and arid, grassy tops. Building as it did on Darwin's geologic observations, Wolf's account became the standard interpretation of island geology.[50]

The United States joined in the hunt for Galápagos specimens with official visits in 1888 and 1891 by the *Albatross*, a state-of-the-art Fish Commission steamer staffed by professional naturalists. Alexander Agassiz, successor to his father at Harvard's Museum of Comparative Zoology, led the second of these expeditions; what he said about it applied to both. "The object," he wrote, "was deep sea dredging, and [it] only included an incidental visit of a few days to the Galápagos."[51] The expedition thus mirrored his father's *Hassler* voyage, though now the dredging equipment worked.

Despite limited objectives, the *Albatross* expeditions marked the formal entry of the United States as a major player in Galápagos science, a position it never relinquished.[52] In 1881, Smithsonian Institution ornithologist Robert Ridgway published a wish list of Central and South American bird species not represented in the institution's museum collection.[53] Apparently reasoning that the United States National Museum should possess a complete set of specimens from throughout the Western Hemisphere, as if to establish its dominion, Ridgway urged collectors to supply the missing items. Some of these species, including several famous from Darwin's work, came solely from the Galápagos. The *Albatross* expeditions offered an opportunity for government scientists to fill this void and to obtain giant tortoises—another hemispheric trophy—for museums and zoos in the United States. Conveniently, Ridgway's boss at the Smithsonian, ornithologist Spencer Baird, also headed the U.S. Fish Commission. With the stroke of a pen, he included a collecting stop at the Galápagos on the *Albatross*'s itinerary, and sent along the Smithsonian's wish list of birds and other specimens.[54]

The *Albatross* stayed in the archipelago for less than two weeks in 1888, but managed to stop at eight different islands. The ship's naturalists and crew gathered specimens at each anchorage, concentrating on birds, reptiles and fish. The landfalls included the abandoned settlement on Charles Island, where the ship's captain reported "great numbers of cattle, horses, mules, donkeys, sheep and hogs were running wild" where native species once flourished, and a new sugarcane plantation and cattle ranch on Chatham Island that also encroached on wild habitat. Of course the ship stopped at Duncan Island for tortoises, with the crew nabbing ten. The Americans also bought another eight tortoises from the proprietor of the Chatham plantation.[55] As Ridgway hoped, however, and Baird directed, the great haul came in birds. "The collection of birds from the Galápagos archipelago is of special interest for the reason that two islands are represented upon which no collections have previously been made," Ridgway exulted, "while other islands have been more carefully explored, thereby adding very materially to our knowledge of the remarkable endemic bird-fauna of these remote and highly interesting islands." He then quoted Salvin's mantra: "The ground is classic ground."[56]

The *Albatross* made fewer stops in the Galápagos in 1891 than in 1888. Baird had died by this time, which perhaps made a difference in the expedition's itinerary. Naturalists disembarked only at the sites of the Charles and Chatham Island settlements, where they collected birds, and on Duncan to gather tortoises and lizards for a Galápagos Island exhibit in the new Pacific Exhibition Room of Agassiz's Museum of Comparative Zoology. In twitting Darwin, the younger Agassiz picked up where his father left off. "Arriving as we did," he wrote, "at the beginning of a remarkably early rainy season, I could not help contrasting the green appearance of the slopes . . . to the descriptions given of them by Darwin." Indeed, based on what little he saw of the islands in 1891, Agassiz deemed them "as favorably situated for cultivation" as Hawaii. Like father, like son—the younger Agassiz also noted that native birds still had not evolved the least fear of humans. For how they now acted, he jabbed, "I need only refer to Darwin's account."[57]

The birds collected on both *Albatross* expeditions went to the Smithsonian for Ridgway to identify. By this time, he had risen to the top ranks in American ornithology and later became the last scientist without a formal education elected to the elite National Academy of Science. Inspired by his Quaker parents from Mansfield, Ohio, with a near-religious love of nature, Ridgway began as a child to send his sketches of midwestern birds to the Smithsonian for identification. The drawings caught the eye of Spencer Baird, who in 1867 asked the seventeen-year-old Ridgway to join the government's ambitious cross-country scientific survey of the Fortieth Parallel, directed by the esteemed Clarence King. Although Ridgway thus cut his eyeteeth as a professional naturalist in the field, he never liked to travel and soon happily settled into a desk job at the Smithsonian, where he served as curator of birds from 1874 until his death more than 50 years later. From this strategic position, the quiet and self-effacing naturalist could get the first look at the bird specimens then flooding into the National Museum from throughout the Western Hemisphere. "During his lifetime he described far more new genera, species, and subspecies of American birds than any other ornithologist," biologist Ernst Mayr notes about Ridgway. Yet among his 550 scientific publications, including ten on Galápagos birds, "there is an almost total absence of generalized or philosophical papers."[58]

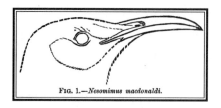

Fig. 1.—*Nesomimus macdonaldi.*

Hood Island mockingbird, from Robert Ridgway's treatise (1889)
(Source: Used by permission of UGA.)

Ridgway's work on Galápagos birds stands out for the sheer number of new species that he described. No ornithologist before or since has named so many new types from the archipelago. In large part he achieved this feat through hard work, as the National Museum's curator of birds at a time when the country's scientific research program took flight. As a result of his position, novel specimens streamed to him in bottles and crates from the Galápagos and elsewhere, each typically tagged with its precise collection site. He also inflated his numbers through an inclination to split specimens into the narrowest of species based on the slightest of differences, especially for specimens collected at different sites—such as on separate islands. In the eternal battle between "lumpers" and "splitters" of species, Ridgway was the ultimate splitter, an approach that became most pronounced with island birds. Later, many of Ridgway's genera were collapsed into species, and his species into varieties.[59]

Even more than was his custom, Ridgway's tendency to split species marked his work on Galápagos finches—particularly the genus *Geospiza*. Darwin wondered whether the obviously similar finches of the various islands constituted distinct species (thus serving as evidence of transmutation) or simply well-marked varieties (thus signifying nothing). Ridgway muddied the waters by identifying so many more finch species. "It will doubtless seem to some that I have gone to an undesirable if not reprehensible extreme in naming so many forms," he conceded in his definitive treatise on Galápagos birds. Yet "whenever there seemed to be a well-defined average difference between specimens from different islands, I have not hesitated to separate them as local forms. No other course, indeed, is practicable; for were 'lumping' once begun there could be no end to it." Indeed, lumping them "might easily end in the recognition of a single variable species, equivalent in its limits to the genus." Thus, he asserted, "the real promoter of chaos and enemy of order is the 'lumper,' and not his much maligned co-worker, the 'hair-splitter.'"[60]

Despite his readiness to split species, which inevitably supported an evolutionary view of their origins, Ridgway hesitated to take sides in the ongoing controversy over how evolution worked. In this he followed his mentor, Baird. Although both of these ornithologists apparently adopted a neo-Lamarckian viewpoint regarding origins (and abandoned their childhood religious beliefs in the process), neither publicized his opinions. "Although our knowledge of the bird life of this interesting island group has been vastly increased since the publication of Darwin's discoveries there," Ridgway wrote in his treatise on Galápagos birds, "the information which has accumulated is still too fragmentary to warrant any serious attempt to solve the problems to which Mr. Darwin first called attention." Instead, the cautious curator characteristically called for more research. "Not a single island of the group can be said to have been exhaustively explored, and few of the species are known in all their various phases," he stressed. "Until all these present mysteries are solved, theories and generalizations are necessarily futile."[61]

This admonition was aimed directly at one particularly dogmatic scientist, George Baur. He and Ridgway represented polar opposites, and their encounter over Galápagos birds inevitably led to bitter conflict. While a museum curator's conservative compulsion to acquire and identify led Ridgway to do his Galápagos research from a desk in Washington, Baur leaped into the field to prove a radical idea. In a controversial 1887 book, the Austrian paleontologist Melchior Neumayr had popularized the concept of land bridges and lost continents to account for the distribution of species in the Southern Hemisphere: connections between continents helped to explain how similar land animals evolved in Africa and South America within the past 100,000 years—the estimated time available. In a flash of insight two years later, the German-born and educated Baur, then working for the renowned paleontologist O. C. Marsh at Yale, saw this idea as an explanation for the Galápagos fauna as well. Indeed, it addressed Marsh's concerns about the rate of evolution. Marsh had struggled for years to shoehorn his fossil finds into a geologic history too short for evolution by random variation and natural selection by envisioning neo-Lamarckian and orthogenetic shortcuts.

"It was by an accident that my attention was directed to a study of the Galápagos Islands," Baur later explained. While unpacking a fossil tortoise from Nebraska for the Yale museum in 1889, he noticed its resemblance to living tortoises from the Galápagos. To the impulsive Baur, this immediately suggested a former land connection between the islands and the American mainland. "In the evening of the same day I wrote in my diary: 'What is the origin of the Galápagos fauna?'," he recalled. "It is not introduced, but left there; the Galápagos originated through subsidence of a larger area of land; they do not represent oceanic islands, as generally believed, but are continental islands."[62]

Baur's idea flew in the face of established Darwinist thinking. No less an authority on Darwinian evolution than Alfred Russel Wallace had recently published *Island Life*, in which he used the Galápagos Archipelago to illustrate the genesis of new species through natural selection of a few original types introduced onto isolated, oceanic islands. Those few types—chance arrivals from the American mainland—evolved to fill available niches on various islands in the absence of other types that did not reach the remote archipelago. "On the whole," Wallace concluded, "we have no difficulty in explaining the probable origin of the flora and fauna of the Galápagos." A land bridge would utterly disrupt his thinking and raise questions about the absence of so many basic mainland types.[63]

Baur disagreed with Wallace. "Tortoises," he countered, "are unable to swim." Even if one managed to reach the Galápagos, "alone it could not propagate. This was only possible after a similar accident imported another specimen of *the same species*, of *the other sex*, to *the same island*." Moreover, the improbable coincidence would have to repeat itself for each distinct island species. Accounting for current patterns of distribution with bridges could more logically explain all the archipelago's land species, Baur maintained. Otherwise, to get similar types to many different islands, "we would have to invoke a thousand accidents." To Baur, the archipelago must be the mountaintops of a subsided continental landmass, and its distinctive fauna and flora the trapped remnants of the former continent's animals and plants. "Every island developed, in the course of long periods, its peculiar races, because the conditions on these different islands were not absolutely identical," Baur postulated, with a neo-Lamarckian bow to the role of the environment in pushing evolution. He knew that the great weight of scientific authority stood against him. Indeed, he all but boasted,

"This result is in direct opposition to the opinion of all authors who have worked on this group of islands." Baur liked it that way.[64]

Determined to prove his theory through fieldwork, Baur quit his job at Yale to explore the Galápagos. To raise funds for a private expedition, he broadened the issue at stake beyond his idiosyncratic theory of subsidence to the fundamental issue of whether organisms evolve in response to environmental stimulus (as the neo-Lamarckians maintained) or whether the environment simply selects beneficial inborn variations (as classical Darwinism held). "There is no other place on the whole earth," Baur asserted, "which affords better opportunities for such a work than the Galápagos. Here we have the original natural conditions, hardly influenced by man. If all the variations of the forms on this group of islands, or even only the variations of a few genera, are studied, and the conditions of each variation examined, then we may perhaps be able to express a more definite opinion on the causes of variation itself."[65]

Baur never explained just how his Galápagos research could resolve this scientific controversy, but he did call for haste due to the threatened status of island species. "Such work ought to be done *before it is too late*. I repeat, before it is too late! Or it may happen that the natural history of the Galápagos will be lost," he exclaimed, "lost forever, irreparably!" Despite a curt refusal of support from his home government in Germany, such pleas raised sufficient funds from private sources in America (including the independently wealthy neo-Lamarckian paleontologist Henry Fairfield Osborn) to pay for Baur and one assistant to travel by commercial vessels and rented sloop to twelve different islands in the group during two months in 1891.[66]

During their time in the Galápagos, the energetic Baur and his assistant collected more specimens than any previous visitors—over a thousand birds either skinned or bottled; hundreds of reptiles, including live tortoises; and an impressive array of spiders, shells, insects and plants. Baur saw in these specimens precisely what he was looking for: distinctive genera spanning the archipelago with island-specific species. Skipping over the question of what caused these local variations (which was wholly beyond the reach of his research program), Baur used his findings to push his pet idea about subsidence. *"How can we explain the harmony of distribution by the theory of elevation?*" he asked in his typically pugnacious style. "The theory of subsidence, however, explains every point in an absolutely easy manner. All islands were connected together at a former period; at this time the number of species must have been small; through [later] isolation . . . every island developed its peculiar races." Some species inhabited more than one island, of course, and some closely related species cohabited the same island, but Baur discounted these observations with a mathematical formula purportedly showing their statistical insignificance.[67]

Most naturalists thought Baur's findings showed a distribution pattern associated with uplifted oceanic islands. Alexander Agassiz ridiculed him for "taking a good deal of poetic licence with our present knowledge." Theodor Wolf added, "Every geologist will stand up against Dr. Baur's hypothesis."[68] Baur saw every such attack as an opportunity to counterattack. His scientific articles became wholly repetitive except for their increasing belligerence.

Baur directed his most vicious assaults against Ridgway, to whom he had entrusted his bird specimens for identification. True to form, Ridgway split Baur's specimens into such narrow species that they did not fit Baur's preconceptions. Worse, in a rare spasm of lumping, Ridgway used the rich array of finches Baur collected to collapse two genera (*Cactornis* and *Geospiza*) into one, which disrupted Baur's predicted pattern of island-specific distribution. Baur responded by publicly and privately accusing Ridgway of various professional sins from dishonesty to incompetence. The mild-mannered Ridgway replied in an 1897 letter to the *American Naturalist*, "To Dr. Baur *Cactornis* and *Geospiza* seem to be distinct, and to have them so would better fit his theory of distribution. To me they are not distinct, because it is impossible to draw any line between them. It is, of course, disappointing to find sometimes that facts do not entirely support our theories." Still in his thirties and planning a second trip to the Galápagos, Baur died of mental exhaustion attributed to overwork shortly after Ridgway's reply. Having lost its most energetic supporter, the subsidence theory of Galápagos origins slowly sank from view.[69]

The lure of giant tortoises drew a second eccentric naturalist into the web of Galápagos science—Walter Rothschild. As a seven-year-old boy born of a union between members of the British and German branches of the richest family of international bankers in history, Rothschild began collecting birds and butterflies at his family's English country estate, Tring Park, in 1878. This first love grew into an all-consuming passion for the brilliant but emotionally troubled Rothschild, who eventually amassed at Tring Park the largest private collection of animal specimens in the world—over 2 million items, mostly stuffed birds and mounted butterflies but also an even gross of live giant tortoises.

Rothschild first became involved with tortoises in 1881, at age thirteen, after meeting Albert Günther at the British Museum. Günther naturally took time for the child because the boy's father—Britain's first Jewish peer, the first Lord Rothschild, whom the press dubbed the "real ruler of England"—was a museum trustee. Günther was genuinely impressed, however, with the boy's knowledge of natural history and wanted to nurture it. This included introducing the lad to giant tortoises as threatened living fossils that held the key to the scientific puzzle of origins. One meeting led to another, and soon the attentive curator became the lonely child's mentor, if anyone could mentor such a secretive and distrustful boy. When Günther died thirty-five years later, Rothschild wrote, "It is as if I had lost a parent."[70]

If Günther became a father figure for Rothschild, then giant tortoises substituted for children as objects of affection. A favorite niece described Rothschild's emotional bond to "his giants" as a "protracted love affair." Although "obviously fond of giant tortoises," the niece explained, "Walter was not, in the early days when Günther first introduced him to these animals, deeply interested in their systematics, classification and relationships—he just cherished and admired them." She believed that their threatened status "engendered specific emotional reactions" in him. "I wanted to save them for science," Rothschild wrote, even if it hastened their extinction in the wild. His was a controlling sort of love, and Rothschild had the means to enforce it.[71]

Rothschild began his affair with tortoises modestly, by trying to get some of them and their remains for Günther and himself. He just never knew when to stop. He first bought them from museums and zoos in Europe—paying any price demanded—and then from collectors in America and elsewhere, including fourteen live ones brought back by Baur, who sold them along with the bulk of his Galápagos collection. Within a dozen years after he began, Rothschild had assembled at Tring Park the largest

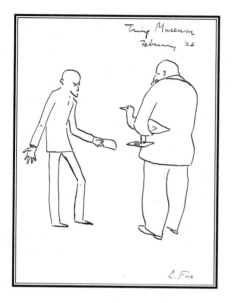

Sketch of Walter Rothschild (holding mounted bird) with Tring Museum ornithologist Ernst Hartert, who coauthored Rothschild's 1899 study of Galápagos birds (1926)
(Source: Reprinted courtesy of Miriam Rothschild.)

herd of giant tortoises outside of the Galápagos Islands, and asked Günther to coauthor with him an authoritative monograph on the animal. Günther reluctantly declined, apparently out of deference to Rothschild's father, who disapproved of his son's preference for natural history over business and politics. The monograph never appeared, but Rothschild continued collecting tortoises and soon started writing technical papers of his own about them—seventeen in all, mostly descriptions of his specimens published in his own Tring Park scientific journal, *Novitates Zoologicae*. Later he leased the entirety of Aldabra Island to protect and collect the surviving Indian Ocean tortoises. The Galápagos Islands remained his principal interest, however, in part because it also appealed to his continuing fancy for birds.

Baur's Galápagos collection came complete with an exceptional array of birds, including those Baur had reclaimed from Ridgway after their falling out. These birds excited Rothschild almost as much as the collection's tortoises. He already possessed the largest private collection of bird skins, eggs and nests in the world; now he added Baur's number to it. "Their study made the desire for more material [from the Galápagos] more ardent," Rothschild wrote. So he started contracting with commercial collectors to gather even more specimens from the archipelago, beginning in

1897 by engaging Frank Blake Webster of Massachusetts. Webster, who operated an international business supplying animals for zoos and museums, chose C. M. Harris to lead the first in a series of Rothschild-funded collecting trips to the Galápagos.[72]

Rothschild's instructions through Webster to Harris identified the cutting edge of scientific collecting activity on the Galápagos. Regarding where to collect, the instructions stated, "If practical you will first visit *Wenman and Culpeper* then the following Narborough, Albemarle, Bindlow, ..." The list proceeded in descending order of interest, roughly reversing the order of previous collecting activity. "Culpeper Never explored," the instructions explained, "Wenman and Narborough the same." Unknown species would most likely lurk in such places. Regarding what to collect, the first and last entry stand out. "YOU WILL CAREFULLY GO OVER THE ENTIRE GROUND OF EACH ISLAND Securing Birds at least 50 of a kind," the initial entry declared. "Note that the slightest difference in bill or size while the bird would in other respects be the same they would be different," the instructions added, in a reference to scientific fascination with the beaks of Galápagos finches. After noting other items for collection "not to the loss of time," the list concluded in large capital letters, "TORTOISE LIVE OR DEAD—Ever[y] specimen that you can obtain large and small." The instructions closed with the weighty charge: "We look to you to out do Expeditions of Darwin, Baur, Agassiz and others. Believing that I have selected a party with NERVE, BACKBONE AND ENERGY, I am yours very truly."[73]

Harris, at least, possessed these attributes in abundance; others in the party died of yellow fever in Panama en route to the Galápagos. After losing his team of collectors, Harris retreated to San Francisco to recruit replacements and then sailed directly with them to the Galápagos and back—skipping the fever-infested islands off Panama and Columbia, where Rothschild had instructed him to look for transitional types linking Galápagos species with mainland ones. The Webster-Harris expedition, as it was called, returned with over 3,000 bird skins and 65 live tortoises from the archipelago. As to both types of specimens, Rothschild hailed the collection as "by far the largest and best ever amassed on the Galápagos Islands."[74] Thereafter, Rothschild employed Bay Area collectors to work the Galápagos.

Specimens from these trips went to Tring Park, of course, where Rothschild and his assistants cataloged them and described any new finds. His tendency to split species naturally led him to claim many new ones—over 5,000 in his career, with 260 genera, species or subspecies named in his honor.[75] The live tortoises reminded in England, with the survivors moving to the London Zoo after Rothschild's death in 1937. Five years before his death, Rothschild sold most of his bird specimens to New York's American Museum of Natural History—some 280,000 skins in all, including every one from the Galápagos, for about one dollar apiece. He needed funds quickly to pay a blackmailer who threatened to tell his elderly mother (with whom he still lived) that he kept two working-class mistresses in separate London apartments. Apparently his mother already knew, but had never let on to her son. Rothschild was heartbroken by the sale, and the London press denounced the unexplained loss to British science. Thus it was in New York, rather than at Tring Park, that the English ornithologist David Lack used the Rothschild collection to work out the speciation of Galápagos finches; but that did not happen for another half century.[76] In the meantime, the initiative in doing Galápagos science passed to California.

FIVE

CALIFORNIA STAKES
ITS CLAIM

WHEN WALTER ROTHSCHILD'S EXPEDITION resorted to San
Francisco as its operational base in 1897, it heralded a permanent transfer
in the locus of Galápagos research and prefaced a continental shift in
American science. Up to this time, established English and East Coast scientific institutions dominated study of the archipelago: the Royal Society,
the Zoological Society of London, the British Museum, the Smithsonian
Institution and Harvard University, reflecting the larger pattern of a few
North Atlantic nations and their elite institutions projecting power and
influence over Pacific science generally. Accordingly, when Rothschild first
conceived of sending an expedition to the Galápagos, he engaged a Massachusetts man, Frank Webster, whose firm had collected specimens for
the Smithsonian and Harvard. Webster then dispatched four collectors led
by Charles Harris to Panama, where they contracted yellow fever. "Can fit
out best in California," Harris wrote to Webster in Massachusetts as the
party fled north. "So far it has been bad luck, but keep a stiff upper lip &
I will carry this through If I don't Kick." Only Harris survived to carry on
the expedition. In forwarding Harris's letter to Rothschild, Webster explained to the English aristocrat, "By stiff upper lip Expression means 'Do
not get discouraged.'"[1]

Harris stopped his flight in San Francisco, where he found ample resources to reorganize the expedition. Whereas in Panama he could not

drum up any interest in the venture beyond suppliers eager to tap Rothschild's legendary purse, in San Francisco Harris worried about exciting
too much interest. Bay Area naturalists seemed intent on building collections of Pacific species for the University of California, Stanford University and the California Academy of Sciences—and the Galápagos fell
within their sphere of interest. Rothschild demanded as many tortoises as
possible and a massive array of native land birds. No one then knew if
enough of these animals remained to satisfy the covetous Rothschild,
much less multiple expeditions, so Harris struggled to keep his destination
secret while assembling a new team of collectors and securing supplies.
Back in Massachusetts, Webster brooded about possible competition from
California naturalists.[2]

Webster had reason to fear. The object of Rothschild's obsession had
become a matter of speculation in the local press, especially after Harris
began recruiting collectors, including Galen Hull (who could skin a finch
in fourteen minutes), Frederick Drowne (a specialist in preserving marine
mammals) and Rollo Beck (already renowned locally for his knack at
catching birds and later hailed by zoologist Robert Cushman Murphy as
"the most successful worker in this branch of ornithology that the world
has known").[3] None of these men were scientists, but California naturalists knew them well and relied on their collecting skills. Soon researchers
at nearby Stanford University began organizing their own expedition to
the Galápagos—the first overseas venture ever attempted by that new
school. The California Academy of Sciences, working with the University
of California, would not be far behind. Spurred by competition—a powerful force in science—Harris worked around the clock to ready his expedition in only five weeks. "Hurrah [we] are off," he wired to Webster on
June 22, 1897. "Galápagos or Bust."[4] The Stanford expedition followed a
year later. Each trip lasted for nearly a year, much longer than any prior expeditions to the archipelago.

*Signature line from Charles Harris's
letter to Frank Webster upon departure for the Galápagos (1897)
(Source: Reprinted courtesy of
Miriam Rothschild.)*

That the San Francisco Bay Area could provide such a wealth of collectors and competitors for Galápagos research testified to the region's remarkable emergence as a center for science. San Francisco scarcely existed a half century earlier, when California became part of the United States in 1848. The Bay Area's population began exploding a year later with the discovery of gold nearby, and never stopped. By 1850, San Francisco already stood as the twenty-fifth largest city in the country, even though California remained among the least populous of states. San Francisco steadily mounted the ranks of American cities thereafter, reaching ninth place by 1900, with nearly 350,000 residents. California rose along with it, reaching the middle ranks in population by the turn of the century—though it still lagged far behind the major eastern states with top scientific institutions. Indeed, in 1900, the population of California roughly equaled that of Kansas and still trailed that of eight other midwestern farm states, including Missouri by a two-to-one margin. Yet Californians possessed a heady sense of destiny that accepted science as a means for progress and the Pacific as their domain. Whereas turn-of-the-century Midwesterners could only have conceived of the Galápagos in abstract terms (just as they might consider Tibet or Timbuktu), California scientists developed a proprietary interest in the place.

Science came of age rapidly in the Bay Area. In 1853, less than three years after their state joined the Union, some local physicians and civic leaders joined in organizing the California Academy of Sciences (CAS). "An Association has been formed in this city," the group boldly announced, "the objects of which as its name imports, are the collection of specimens and the investigation of all matters pertaining to Natural Science in the State of California and on the Pacific coast, and the adjacent island region."[5] As if to confirm its Pacific reach, the CAS then elected a Hawaiian among its first corresponding members and accepted a donation of marine shells and coral from the South Pacific as its first specimens.[6]

During the 1860s, the Bay Area gained two of the nation's premier naturalists when Joseph Dewight Whitney took over the California Geological Survey and Joseph LeConte became a geology professor at the newly founded University of California. Although neither extended his own research to include the Galápagos Islands, both excited local interest in the scientific issues raised by that archipelago by becoming prominent early proponents of the theory of evolution. Largely due to their influence, California scientists took the lead in embracing the theory, albeit with a neo-Lamarckian flavor.[7] Further, as CAS officers, Whitney and LeConte pushed the academy to expand its research and collecting activities, which ultimately included some of the most important scientific work ever done on the Galápagos.[8]

A brief visit by Louis Agassiz propelled the development of Bay Area science. Having taught LeConte at Harvard and served on the faculty with Whitney there, Agassiz was assured a warm reception when he disembarked in San Francisco from his *Hassler* expedition in 1872, but even that proud man could not have anticipated the public response to his arrival. As the best-known American scientist of the day, Agassiz received a hero's welcome to a city whose leaders fretted about their cultural isolation on the West Coast. Reports of the *Hassler*'s arrival dominated every local newspaper, with one of them devoting its entire front page to a sketch of the famed naturalist. Basking in the acclaim but weakened from his strenuous voyage, Agassiz remained in California for over a month. He delivered two immensely popular public lectures in San Francisco that drew attention to Galápagos research and challenged local entrepreneurs to support science.[9]

His first San Francisco address was at a welcoming reception hosted by the CAS. "You are surrounded with wealth as no State ever was," he admonished members and their guests, "and yet I see [the Academy is] still in close quarters. You have not rooms in which to display your acquisitions," which now included Galápagos specimens just donated by Agassiz. He then related how wealthy Bostonians had financed his Harvard research museum and scientific expeditions. "You are richer than Boston—a great deal more so," Agassiz prodded. "This is what the world wants—not books to read, but men to learn what is not yet known. Those men cannot be educated in the school-room. They must be educated in nature, among specimens." According to Agassiz, scientific research through field expeditions and museum collections offered the key to intellectual and economic progress. Looking around him at the likes of Leland Stanford, Charles Crocker and Mark Hopkins—three local merchants who had recently acquired legendary wealth by building a railroad to the East over seemingly impassable mountains—Agassiz concluded, "What you have achieved in material prosperity, do not despair to achieve in intellectual growth in the same short time." Stanford credited this speech as inspiring his later largesse toward science and education.[10]

California reaped the rewards of Agassiz's orations for a generation. Membership in the academy surged, especially among the city's upper crust. Stanford promptly became a life member and soon joined with Crocker in financing the acquisition of a first-rate display collection that put the CAS on course to assembling one of the country's leading natural history museums. Crocker also endowed an expedition fund for the academy, and his sons joined with real-estate tycoon James Lick to build it a monumental exhibition hall.[11]

When the Stanfords' only child died in 1884, the grieving parents devoted their fortune to creating a scientific research institution in the Bay Area to memorialize the boy—Leland Stanford Junior University. The school began with four times more money than Harvard then possessed and a mission for practical scholarship serving California and its "Pacific empire." To lead his new school, Stanford hand-picked Agassiz-trained marine zoologist David Starr Jordan, who publicly championed the theory of evolution as avidly as his former teacher had attacked it, but (like LeConte) with a non-Darwinian twist. Mark Hopkins's sole heir, Timothy Hopkins, gave the new university a state-of-the-art marine biological laboratory in 1892, modeled on what Louis and Alexander Agassiz had launched for Harvard. He then financed the lab's first overseas scientific expedition, to the Galápagos in 1898–1899. The younger Hopkins also subsidized a much more extensive CAS voyage to the archipelago six years later. Spurred by Stanford University's example, the University of California reorganized itself into a world-class institution that ultimately became a recognized center for Galápagos research.[12]

By 1900, the Bay Area had achieved all that Agassiz had called for a quarter century earlier. Fittingly, the Harvard naturalist's statue towered over Stanford's main entrance, and academy members still repeated Agassiz's admonition to them: "I hold that it is your mission to show to the western part of the continent that without intellectual growth there is no greatness for the State."[13] California opted for greatness.

For a generation, Jordan stood at the center of Bay Area science by presiding over both Stanford and the CAS. He had started his scientific career as a botanist at Cornell University but was inspired to study fish while taking a summer course at Agassiz's marine biology laboratory in 1873. Agassiz impressed on Jordan the importance of fieldwork and studying specimens. "Never try to teach what you do not know yourself and know well," Jordan later recalled Agassiz saying. "Take your texts from the brooks, not from the booksellers." Consequently Jordan participated in as many expeditions as his administrative duties permitted and made his name in science as a taxonomist who identified more than 2,500 species of fish. "Whenever a new lot of fishes from some distant place had been spread out on table in the Stanford zoology laboratory, the president would be given a call," one official university history notes. "He would stride up and down like an officer reviewing troops, and, unerringly, he would stop before a species new to science. 'Ho,' he would exclaim in his great sonorous voice, 'that's a new face!'"[14]

As a prominent evolutionary biologist, Jordan naturally wanted to examine specimens from the Galápagos and, although he never went to the archipelago himself, he made sure that Stanford and CAS collectors did so. Jordan's interest in Galápagos species went deeper than a mere sentimental attachment to the place where Darwin had "discovered" evolution: it fit a larger research agenda. At the time, American and European scientists uniformly accepted the fact of evolution but greatly disputed its processes. "The days have long passed . . . when the truth or falsity of the law of organic descent is a debatable thesis," Jordan wrote with his Stanford colleague Vernon Kellogg in their classic 1907 treatise *Evolution and Animal Life*. "But in what particular way, or as the effect of what particular causal factors, this descent or transformation of species, that is, kinds of organisms, comes about,—here there is unlimited field for debate and polemic."[15]

On both scientific and philosophical grounds, Jordan objected to natural selection as evolution's principal driving force. Looking around them in the barren American West, California naturalists tended to see adaptation to harsh environmental conditions as a more significant feature in the evolutionary process than competition among species.[16] Many, like LeConte, turned to neo-Lamarckian theories involving the inheritance of individual acquired characteristics. By the turn of the century, however, Darwinists had fairly well discredited such notions—at least in Jordan's mind. Yet as a devout pacifist and future leader of the international peace movement, Jordan would not return to a "struggle for existence" alone as the source of life's novelty and variety.[17]

Like many great scientists, Jordan never separated his science from his religion or philosophy.[18] In this respect he followed his mentor, Louis Agassiz, and his senior colleague Joseph LeConte. For all their differences over the theory of evolution, Agassiz and his devoted students LeConte and Jordan shared a sense of the divine in nature. Although they eschewed traditional Christianity, each was religious in his own way. Agassiz saw an immanent God in nature successively creating species; LeConte's God worked through indwelling Lamarckian forces; while for Jordan, nature itself drove the evolutionary process—but never in a purely materialistic sense. As befit the leader of a dynamic scientific and cultural community isolated in the still-remote American West, Jordan turned to isolation as a causal factor in evolution, making its examination his major theoretical contribution to science.[19] As a natural laboratory for studying the effects of isolation on species formation, the Galápagos could possess profound beauty for someone like Jordan, just it had for the Agassizes. Here he could all but touch his God.[20]

Elaborating on his view of evolution, Jordan once explained that "in Darwin's view, isolation or segregation was doubtless a feature of Natural Selection, not to be set off against the latter as a separate factor in descent." This was not enough for Jordan. "Among the factors everywhere and inevitably connected with the course of descent of any species, variation, heredity, selection, and isolation must appear," he stressed. Of these, however, "isolation as a factor longest overlooked, though to the field naturalist the most conspicuous of the four, must be advanced to the post of honor beside the others." Simply isolate a population by geographical or other barriers, Jordan all but declared, and through variation, heredity and selection a new species will naturally evolve.[21]

More a change of emphasis than an alternative explanation for evolution, Jordan's approach nevertheless allowed the pacific naturalist to fasten on something other than the hateful struggle for survival. From his work with isolation as a factor in evolution, he developed Jordan's Law: that the species most closely related to another is found just beyond a barrier to distribution. Jordan focused his attention and that of his students and colleagues on similar species living across slight geographical barriers, such as finches living on the various Galápagos Islands. He featured these finches in popular writings to illustrate the impact of isolation on the origin of species but gradually lost interest in them as the teams he sent to the Galápagos found (contrary to the received wisdom) that various species did not neatly segregate onto separate islands. Indeed, some of them freely flocked together. Decades later, selection theory later would return to the forefront of evolutionary science when it succeeded in explaining this phenomenon. At the time, the archipelago did not appear to offer Jordan the marine biologist (as opposed to Jordan the popular writer) a particularly good laboratory for researching the impact of isolation, because the barren islands did not harbor many isolated fish pools. Thus he directed his own research elsewhere and sent ornithologists, herpetologists and botanists to the Galápagos—though he closely followed their findings.[22]

Prior to the opening of the Panama Canal in 1914, the archipelago was half the world away from Europe and the East Coast of the United States but could be reached in a matter of weeks by ship from California. San Francisco–based vessels transporting cargo to and from the west coast of South America often called on settlements in the Galápagos to stock up on seals and tortoises.[23] Years of experience made the archipelago a familiar spot for San Francisco sailors. Harris discovered this when reorganizing his expedition for Rothschild in 1897, and Stanford University relied on it a year later when arranging transport for its explorers. As the Bay Area scientific community came of age, naturalists at Stanford and the CAS began viewing the Pacific Ocean as their proper domain and the Galápagos Islands as its prize jewel. Further, California naturalists and collectors, accustomed to desert offshore islands and stark mountains, were more at home in the Galápagos than their English and eastern counterparts who resided amid lush woodlots and the tangled bank.

Seven scientific expeditions, all from the Bay Area, went to the Galápagos between 1897 and 1906—more than in any other decade before or after, until the establishment of a permanent research station on the islands in 1964. This sudden burst of activity reflected a confluence of factors.

The heightened scientific interest in the archipelago stemmed largely from George Baur's work there in 1891. Seeking to prove his idiosyncratic notion of continental subsidence, Baur had gone to the Galápagos looking

The Kicker Rock.   400 feet high.

Sketch of Kicker Rock near Chatham Island, from Charles Darwin,
Geological Observations *(1897 ed.)*
(Source: Used by permission of UGA.)

for a "harmonic distribution" of closely related species on the various islands. Where else could Galápagos animals come from, especially those
like the giant tortoises and ground finches that could not swim or fly far,
if they were not surviving species clinging to the remnants of a sunken
landmass? What else could account for the striking similarities among
them?[24] Baur did not win many converts to his subsidence theory, but scientists were impressed by the stunning array of specimens that he and his
assistant, C. F. Adams, brought back from the Galápagos—including 1,100
bird skins and 21 live tortoises.[25]

For the first time, naturalists had ample series of Galápagos finches and
tortoises for making evolutionary classifications. Baur first loaned his birds
to Robert Ridgway for identification and then sold them to Rothschild,
while his surviving tortoises ultimately went to Rothschild. These naturalists split Baur's specimens into a multiplicity of species based on ever so
slight differences in average physical characteristics. Many of these divi-
sions did not stand up over time. Rothschild, for instance, divided Galápagos tortoises into eleven species, where today all are lumped into one. At
the time, however, this multiplicity of species created the impression that
the seemingly barren Galápagos Islands harbored a wealth of novel types

whose survival was threatened by introduced competitors. "Save them for science" became the cry, even if it meant capturing or killing the last one in the wild.[26] Rothschild and the Californians were equally determined to play the hero's role in this rescue effort. The resulting competition for specimens drove Galápagos science for a decade.

The turn-of-the-century Galápagos expeditions were all fundamentally similar, not least because none of them included any scientists. Just as the established naturalists Sedgwick and Henslow remained behind at Cambridge in 1835, and sent their student Darwin on the *Beagle* to do the collecting, so these later efforts relied exclusively on paid collectors, biology students and museum staff. The trips took time because evolutionary biologists and research museums required a rich array of specimens to investigate subtle variations within and between species, not just a few representative samples to identify basic types. The Webster-Harris and Stanford expeditions of 1897 and 1898 each stretched for nearly a year, and the 1905–1906 CAS trip extended over seventeen months. Rothschild could not have left his business and political concerns for that long, and the scientists at Stanford and the CAS had teaching and museum duties. They wanted collectors willing to bring back proper scientific specimens, not necessarily trained naturalists who would do the scientific analysis.

Given their lack of prior expedition experience and formal scientific training, the participants in these expeditions succeeded remarkably well. Rothschild employed young American collectors: Harris, Hull, Drowne and Beck for his first expedition to the Galápagos, then Beck for two more and G. M. Green for one. Through Webster, he also commissioned W. P. Noyes, the captain of the ship that carried the Stanford expedition in 1898–1899, to collect tortoises for him during a 1900 voyage to the archipelago. Beck, Green and Noyes called on sailors and Galápagos settlers for assistance in their efforts. Stanford University sent two of its most promising zoology students, Robert E. Snodgrass and Edmund Heller, both age twenty-three. The California Academy of Sciences picked Beck to lead its 1905–1906 expedition and assembled seven young collectors to assist him: two Stanford undergrads, W. H. Ochsner and F. X. Williams; four junior museum aides, Edward Winslow Gifford and Joseph R. Slevin from the CAS staff, the Smithsonian's Alban Stewart and J. S. Hunter of the University of California; and an energetic Bay Area schoolboy named Ernest S. King to help catch tortoises and iguanas.[27]

None of the participants in these expeditions was older than thirty when they first embarked for the Galápagos, and three were still teenagers. Only Stewart had received any graduate instruction in science, but he held solely a master's degree in botany. Beck, a pious Methodist, apparently did not even believe in the theory of evolution and enforced the Sabbath during the CAS expedition, much to the amusement (and relief) of his largely irreverent, exceedingly overworked assistants.[28] Inspired by their Galápagos experiences, some of these collectors later became respected scientists. Gifford, for one, ultimately became a full professor of biology at the University of California—the last person to attain that rank at an American research institution without ever attending college. Others, such as Beck and Heller, never lost their wanderlust and went on to ever more exotic destinations.[29] Taken together, these seven expeditions provided the specimens that revolutionized scientific understanding of the archipelago and helped rehabilitate natural selection as the driving force of evolution.

Unlike most earlier expeditions to the Galápagos, these turn-of-the-century ventures aimed at more than simply looking for novelties and gathering representative samples. They sought comprehensive research collections suitable for investigating subtle evolutionary relationships. For his part, Rothschild had moved from wanting pet tortoises for his Tring Park menagerie to demanding a complete series of tortoise specimens for his growing Tring Museum research collection, which (aided by Albert Günther) he used in making taxonomic classifications. He did the same with Tring's vast array of bird specimens, regularly announcing the discovery of new species. Rothschild took great offense when scientists challenged his taxonomy. The British Museum ornithologist P. L. Sclater's rejection of a proposed new bird species led him to fume, "I feel *very hurt* at the way he treated my paper . . . . I am not an ignorant amateur." Such slights fed Rothschild's obsession with obtaining complete series of Galápagos land birds.[30] When Stanford's student team succeeded in collecting more types of tortoises than Harris's professional collectors (although fewer total specimens), Rothschild abruptly sent Beck twice more to the archipelago with precise instructions to fill gaps in the Tring collection. Green and Noyes further added to this effort, as did Rothschild's successful bid to acquire eighteen of Stanford's tortoises.[31] Each succeeding expedition enriched Rothschild's collection of Galápagos Tortoises and land birds.

Stanford University also saw an opportunity to build its scientific reputation on the back of Galápagos research, with Timothy Hopkins footing the bill for a voyage in 1898. Jordan grandly called it the Hopkins Stanford Galápagos Expedition, even though only two students were involved. Stanford zoology professor Charles H. Gilbert instructed the students simply to go everywhere in the archipelago and collect everything. They did their best. "The value of the large collection which these students brought home with them lies in the thoroughness with which the seventeen main islands of the group were gone over," a Stanford publication boasted. "Since Darwin's visit no such valuable collection has been brought from this part of the world."[32] The expedition's haul included 1,200 reptile specimens, 26 of them giant tortoises; thousands of insects and shells; hundreds of birds, spiders and echinoderms; and 23 new species of fish identified by Jordan himself. Several faculty members joined in published articles about the Galápagos specimens, which became the core of Stanford's biology museum.[33]

In terms of institution building (or rebuilding, as it turned out), the Rothschild and Stanford expeditions paled in comparison with the 1905–1906 CAS venture. The academy already boasted the largest assortment of natural history specimens on the West Coast, featuring superb research collections of California species housed above a popular exhibition hall in downtown San Francisco. Academy director Leverett Mills Loomis and his well-heeled board of trustees conceived of the Galápagos expedition as the first in a series of ventures to expand the research and display collections to encompass the whole Pacific. "The academy realizes that California holds the Golden Gate, not only for purposes of commercial expansion, but also for scientific exploration and research," a *San Francisco Chronicle* article explained in jingoistic terms typical of the day. Building on the academy's various California collections, the newspaper reported, "the forthcoming expedition is commissioned to achieve for all the sections a similar renown as regards to island biology," that is, "to make these study collections so complete as to be in certain respects unique, unsurpassed and unsurpassable."[34]

The sheer magnitude of the CAS expedition dwarfed the others. It was the single most comprehensive natural history survey of the Galápagos ever conducted: eight collectors working in the archipelago full time for more than a year at twenty-nine separate locations on twenty-three different islands. Charged with gathering a representative array of everything that lived in the archipelago, the CAS explorers succeeded in retrieving over 75,000 specimens, more than all previous expeditions to the Galápagos combined. This number included 264 tortoises, which far surpassed Rothschild's take and came after the English aristocrat thought his collectors had cleaned out the population on most of the islands. When the CAS explorers learned, midway through their stay, that a great earthquake and fire had destroyed downtown San Francisco, including their natural history museum, one of them reportedly declared, "We are the collection of the Academy now!" Rather than return home immediately, they continued collecting.[35]

Curiously, the specific object of investigating the theory of evolution rarely appeared in the public press or the private journals of participants. Official CAS records never mentioned it either. Even Darwin received scant mention, except as a famous previous visitor to the archipelago—"It's Columbus," as one San Francisco newspaper erroneously identified him.[36] "The islands have long been known to scientists the world over because of their richness in plant and animal life," a typical press account noted. "Now much of the fauna and flora of the islands is rapidly becoming extinct, and it is with the purpose to gather specimens of these rare plants and animals that the expedition has been fitted out."[37] Describing the Galápagos as a "wondrous dying world," a feature in the *San Francisco Chronicle* noted that "every island will be visited by the explorers, and each island will furnish its individual collection, so that the many scientific problems of geographical distribution, local variations, etc. may be duly studied by the savants of all nations who will repair to San Francisco for this end." These problems interested scientists because they were relevant for understanding the evolutionary process, but such matters seemed highly esoteric for a museum expedition. At a time when conservationist John Muir dominated the Bay Area scientific scene with his highly publicized efforts to preserve wilderness, the CAS promoted this trip as a daring adventure to collect threatened "tropical treasures from the south seas."[38]

Events surrounding the CAS expedition's departure reflected the public's keen interest in such efforts. California citizens subscribed funds so that the CAS could buy and refit a surplus U.S. Coast Survey schooner for its expedition. A large crowd turned out for the ceremonies as the two-masted sailing ship, rechristened the *Academy*, prepared for its departure. "Those assembled occupied every available space on the deck and the cabin," one observer wrote, with the crowd "composed of San Francisco's most prominent men and women."[39] A poem composed for the occasion captured the spirit of adventure then kindled by exotic scientific expeditions:

> OH it's HO for the Galápagos
> Far southward o'er the sea,
> Whose lonely caves' low booming roar
> Breathes forth a mystery. . . .
> Where monstrous turtles crawl about
> All sizzling in the sun,
> Until the cook in strident tones
> Cries, "Fetch him in he's done." . . .
> To you most gallant mariners
> That boldly brave the main,
> Your daring in this enterprise
> Will in our hearts remain.[40]

The *Academy*'s voyage roughly coincided with the sensational polar expeditions of Peary, Amundsen and Scott. To the general public, the Galápagos seemed as mysterious and daring a destination.

Even departing from San Francisco, any extended trip to the Galápagos posed severe logistical problems in 1905. Despite the advent of steamships and their use for brief stopovers at the archipelago, these expeditions used sailing ships because the Galápagos offered no place to refuel, and shallow-draft, island-hopping vessels could not carry enough coal or oil for prolonged cruising. This left the explorers at the mercy of the region's notorious calms and currents. For example, what should have been a quick trip between the southern Galápagos islands of Charles and Hood stretched into a three-week ordeal for the CAS expedition during the dog days of 1906. "Still at sea south of the group; sailing around and fighting with our apology for a navigator," read the entry for ten days running in the *Academy*'s log. "Light winds and strong currents keeping us from making good time." The next three days grew worse: "Dead calm now and drifting further south."[41]

The Webster-Harris expedition of 1897 encountered similar delays and suffered the added hindrance of a captain more concerned about his vessel than Rothschild's collection. "Calm again today," Harris's journal noted as the daytime doldrums set in toward the trip's end. "In the night the captain won't get nearer than 6 to 8 miles from the island; and before he can get to the place in the morning the wind is gone." As the calms spread during the ensuing weeks, Harris ultimately cut short the trip. "Main reason," he wrote in a final journal entry, "if becalmed for a while we should lose the [live] tortoises, run short of food, and have trouble with the captain and crew; so in the morning we will sail for California." The return voyage took six grueling weeks in light winds, with Harris arriving in a state of nervous exhaustion bordering on collapse. "The strain of the last eight months has evidently been very hard upon him," Webster was told.[42] The *Academy*'s trip home took even longer, dooming most of the tortoises. It ended only when the becalmed schooner got a tow for the final stretch.[43]

Travel for these expeditions was difficult even in the best of times. The CAS's *Academy* and Rothschild's leased *Lila and Mattie* carried no electronic communication or navigation equipment and relied on the sun, stars and dead reckoning to find their way. As a result, they occasionally got lost. The other expeditions lacked even the luxury of their own ships and had to go where they could find passage. Further, the seventeen-month CAS expedition packed a scientific staff of eight young men and a crew of three into a ship only 89 feet long and 23 feet in beam. The *Lila and Mattie* was slightly larger but added a crew of five. Both vessels were overloaded with supplies on the outbound voyage and specimens on the return trip, including 264 living, dead or dying giant tortoises on the *Academy*. "The deck is crowded with them and we are skinning them as fast as possible," read a typical journal entry from that return voyage. "As soon as [the skinned carcasses] soak for a couple days we shall be able to knock down the pickle tubs and stow them away." Both vessels also became thoroughly infested with flies, fleas and bugs. "Opened up the hatch at 8:00 AM," the

Academy's log recorded after a failed attempt at fumigation with smoldering sulphur. "Although there were many dead cockroaches and flies scattered about, the bedbugs were as good as new. We had breakfast at 10:00 AM in the sulphur-scented cabin."[44]

The islands offered little respite from shipboard discomforts. Coastal areas were so rocky and the highlands so wet and infested with mosquitos that the explorers slept on board except during overnight foraging or collecting trips into the interior of the larger islands. "Spent the night on shore and tried to get a pig," J. S. Hunter complained in his diary. "The only thing that I did get were about a million mosquitoes. Between them and watching for a pig I got about half an hours sleep."[45] Only the Stanford team established island bases, which they used when the *Julia E. Whalen* left them at different collecting sites.

At the time, the Galápagos had only three settlements. The largest was the administrative capital on Chatham Island with a few hundred residents, monthly boat service to the mainland and nearby agricultural plantations. "There are large sugar-cane fields on both sides," Harris noted in his journal. "Lemon, orange and fig trees line the roadway." After months clinkerbound on barren islands, Drowne declared that the vista "fulfilled my vision of a tropical paradise." Here the explorers sent and received mail and occasionally obtained fresh fruit, drinking water and other supplies. The Chatham settlement also served as the operational base for the three solo expeditions.[46] Smaller settlements on Charles Island and south Albemarle provided local guides and some relief from monotony but little else.[47] Except for a few chance encounters with Ecuadorian tortoise hunters, the explorers otherwise remained isolated.[48]

The island's native barrenness forced the expeditions to carry their own food and water.[49] The *Academy* and *Lila and Mattie* were stocked with seemingly endless supplies of canned California salmon and fruit, dried beans and hardtack. These provisions served as the primary food on the voyage out and back—so much so that some of the explorers vowed never to eat those items again. "The cook parboiled and fried an albatross for supper this evening, some of the party wanted to get the taste of salmon out their mouths," the *Academy's* log recorded at one point. "The bird did not prove a grand success." To supplement their diet in the Galápagos, the explorers devoted inordinate time to hunting feral pigs, cattle and goats.

"We always lived on tortoise liver when obtainable," the CAS's herpetologist added, it being a salvageable by-product from skinned specimens.[50]

As bad as conditions were aboard ship, they were worse ashore. Charged with collecting natural history specimens, particularly tortoises and birds, the explorers had to go where those animals lived, which often meant up mountains, over lava fields and into lagoons. There simply are no pleasant places to go on foot in the Galápagos. The laments leap from every page of the explorers' diaries. "The flamingoes were wading in a slimy ooze," Drowne observed at one point. "When shot we had to wade into this ooze to recover them. In one case Beck got in up to the breasts— a very disagreeable business."[51] There was not enough fresh water on the Galápagos to wash, ever. Harris described Bindloe Island as "an immense lava bed, crowned with a few hills," and wrote of "wrapping [his] feet and legs in canvas" to cross it. "It is terrible getting about here," he added about Albemarle.[52] After struggling to collect specimens on similarly volcanic terrain, Slevin complained, "Feet burned and blistered from yesterday's hunt for turtles. By putting on a plentiful application of cheese cloth, Vaseline and cocoa butter, I got on my shoes and went ashore with the boat early in the morning. This time for lizards."[53]

Sometimes the explorers simply gave up. For example, the *Academy*'s log reported near its end, "We found Culpepper [Island] a most uninviting spot, being merely a flat-topped rock." Birds nested at its top but sharks circled below and a sheer rock cliff rose from the narrow, boulder-strewn shore. "Finding the summit quite inaccessible, we gathered a few sea iguanas and shoved off."[54]

For the *Academy* expedition, rock bottom came after a day collecting on Albemarle when they tried to load two large tortoises onto one small skiff to ferry them several miles around the coast to the schooner. "During this procedure our skiff turned broadside on to the swell and, an extra heavy roller coming in, the skiff capsized, throwing both tortoises, the oars, and the remaining contents of the skiff overboard," Slevin reported. "We tried to pull the boat along the rocks to the beach but the swell was so heavy it smashed into a thousand pieces." Beck had lost his shoes and Slevin his shirt; now the entire party had to travel on foot over razor-sharp lava boulders for three hours to reach their ship's anchorage. "As we walked through the brush in the dark," Slevin wrote, "I felt as if there was not a cactus or

Sketch of Duncan Island noting tortoise habitat, from Frederick Drowne's diary
in Novitates Zoologicae *(1899)*
(Source: Used by permission of UGA.)

thorn bush on all Albemarle Island that I missed running into. However,
I wouldn't have traded places with Beck for anything." The ordeal had a
surprise ending. The next day, the *Academy's* log noted, "Williams, happening to go on deck, sighted one of the tortoises we lost yesterday drifting down past the anchorage. It was bobbing about like a cork, its long
neck protruding far out of the water." Then came the other. "The tortoises
had been in the seawater about 18 hours and seemed none the worse for
it," Slevin remarked following their recapture at sea. The beasts could not
swim, he observed, but perhaps their ability to float in the ocean explained
how their ancestors first arrived on the Galápagos.[55]

The challenge of catching and carrying giant tortoises made these expeditions hellish work. The bigger ones could weigh up to a quarter ton.
Those that remained after two centuries of tortoise hunting tended to live
in the least accessible reaches of the least inviting islands. Yet collecting
them was the chief object of all the turn-of-the-century expeditions.[56]
Harris called it "the hardest work that I ever did for my part, and I guess
the rest thought the same." He related one particularly grueling struggle
to extract eight living giants (Rothschild wanted them alive) from a crater
atop Duncan Island. "It was hard getting them up the side of the crater,
walking being so rough and thorns so plentiful. But this was nothing to be
compared with going down on the other side, which was very steep and
terrible walking." One of the men wrote of "tumbling over lava blocks,

tearing through thorn bushes" to reach a 75-foot cliff, from which the tortoises were lowered by ropes into a skiff. Six of the eight survived.[57]

The Herculean labors of the *Academy* explorers on Indefatigable exemplified the work generally. The CAS expedition spent nearly two months there because, on a previous expedition, Beck had discovered a few survivors of a race of particularly large tortoises thought to be extinct. Claiming it as a new species, Rothschild had called it *Testudo porteri* in honor of the American sea captain who had first reported the differences distinguishing Galápagos tortoises.[58] Now Beck wanted more samples of this giant among giants. The explorers found good anchorage on the island's south coast in a cove that they named Academy Bay. They chose well: it later became the main port for island tourism and site for the archipelago's permanent research station, but at the time it afforded little more than some fresh water "not to be recommended except as a last resort" and access to the interior "through rough country overgrown with cacti."[59]

"On landing and hauling up the skiffs beyond the high tide line, all hands headed inland towards the base of a hill about two miles off, as in this direction appeared the best looking tortoise country," Slevin recorded in the *Academy*'s log. After a day spent scouring the rough countryside, they found two tortoises far too large to carry out alive, one male and one female. "The latter we killed and took out the liver to bring back for supper. The male tortoise was turned over on its back and all four legs stretched out and made fast with lashings to the nearby trees so he would not travel inland and make us pack him further the following day." Returning the next morning, the explorers skinned both tortoises. "They were then lashed on poles with a blanket wrapped around each end of the pole so that they could be carried on the shoulder," Slevin reported. "With two men to a tortoise, we began the journey to the landing place over rough lava and through heavy undergrowth, which had to be cut away at times to make way for the packers. The tortoise packing was worked in relays, as it is hard on the shoulders despite the padding of blankets."[60]

The CAS explorers collected more tortoises from more islands than any scientific expedition in history. Tiny Duncan Island yielded nearly a hundred specimens, even though the Webster-Harris expedition had reported taking them all ten years earlier.[61] Mountainous James Island provided the most onerous work. "The country was so rough and hard to get over," Slevin complained in his journal, "it will be impossible for us to get them [all] out. It was something rich to get out the ones we did. No wonder people don't find tortoises on James."[62] The sheer abundance of tortoises remaining on the big island of Albemarle made it the easiest place to gather live specimens near the coast, but that also attracted settlers who harvested the beasts for oil and food. "Beck worked around the south side of the crater and found an old camp where about 70 tortoise skeletons were scattered about," Slevin noted about one frustrating landfall on Albemarle. "We saw a few to the northward [but] no doubt the natives cleaned the tortoise out very throughly." The CAS explorers tried again using guides from the island's small settlement. "We found two [tortoises] near the house but only moderate-sized ones. We killed one to get the liver for lunch and while we were hunting, our native guide slipped out and cut one of the legs off for his own lunch so we only saved the skull," Slevin commented. Wild dogs ate the rest, shell and all. "We started out next morning with two mules and guide to go along the trail where big tortoises used to be abundant but found they had been slaughtered by the wholesale." The CAS team ultimately bought over three dozen tortoises from local settlers to augment their own catch, making Albemarle their single best source for the reptiles.[63]

The *Academy* returned to San Francisco with 264 tortoises from ten different islands. Fifteen of them survived the trip alive, but they soon died in their pens at chilly Golden Gate Park. Most of the animals were stuffed for display. Because the taxidermists used arsenic as a preservative, the entire collection now resides in a locked vault beneath the museum, where researchers still go to study variations among different island types. There is no better collection anywhere.[64]

"For a long time it was supposed that the tortoises inhabiting all the islands of the group were of one kind, but in recent years it has been discovered that each island has its own peculiar species," the CAS boasted in 1906. "On several islands the tortoises were exterminated before this fact was known, and to what species they belonged has been a matter of speculation among herpetologists. The thorough search made by the members of the expedition has resulted in the discovery of two survivors of those tortoises so long supposed to be extinct."[65] One of these survivors was the last tortoise ever seen on Narborough Island. After an intensive two-day search, Beck found this "old male" deep in the island's mountainous interior, too far from the coast to pack out alive. His field notes describe skinning it by moonlight and lugging the giant carcass out at dawn.[66] Although scientists now identify this and the other Galápagos Tortoises as different island-specific races of a single species, the group remains a classic example of adaptive radiation in its evolutionary development. Those from flat, dry islands have saddleback shells allowing them to reach up their heads for scarce vegetation, while those from moist, mountainous islands have dome-shaped shells better suited for pushing through the dense underbrush.[67]

All of the turn-of-the-century expeditions had the secondary objective of collecting birds, especially finches. Thanks to the Darwin legend, the archipelago's finches were as famous among scientists as its tortoises, and the vast collections of them assembled by the Rothschild, CAS and Stanford expeditions played an even greater role in the history of science. These expeditions brought back thousands of other bird specimens as well, all skinned, stuffed and crated on the spot to prevent spoilage. The expedition journals become redundant in describing the process. "Found birds very abundant," Drowne noted in one day's diary entry. "Several kinds of *Geospiza, Nesomimus,* doves, warblers, flycatchers, *Certhidea,* and several species of *Camarhychus.* Water birds were quite numerous. Saw a number of pelicans, little herons, boobies, curlews, tattlers, and oyster-catchers. Hawks were very abundant. . . . Skinned birds all the afternoon."[68] Sticks and stones proved sufficient for bagging most types of land birds. A flock of water fowl could be shot one by one without the others taking flight. They caught some birds in nets, including Harris's "most extraordinary discovery" of a flightless cormorant, which a grateful Rothschild named *Phalacrocorax harrisi.*[69] In many places the ex-

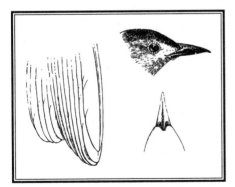

Warbler finch, from Sclater and Salvin's report (1870)
(Source: Used by permission of UGA.)

plorers reported that closely related species mingled freely together, including various types of finches. This drove at least one of the Stanford students to ask, in his field notes, If isolation alone bred species (as Jordan had taught him), then why were these related types together?[70]

Sheer numbers tell the story. The Webster-Harris expedition returned with nearly 3,500 bird skins and eggs, including seven species new to science. Rothschild hailed the haul as "by far the largest and best ever amassed on the Galápagos Islands," but this distinction did not last a year. The Hopkins Stanford expedition bested it, leading Rothschild to commission Beck, Green and the captain of the Hopkins Stanford vessel to fill in the resulting gaps in his own collection. The CAS explorers topped them all, with well over 10,000 bird skins and eggs, again including many new finds. Field notes about the birds came with the specimens.[71]

For the first time there existed two superb research collections—the larger held by the CAS in San Francisco (which ultimately incorporated the Stanford collection), the smaller assembled by Rothschild at Tring—where ornithologists could compare hundreds of preserved specimens of nearly every type of Galápagos bird. Now they could measure the beak of each finch, study every mockingbird's markings and otherwise calibrate differences within and between related species. This is the stuff of evolutionary science, at least so far as it can be done in the lab. "It's impossible to resolve this question" of evolutionary relationships in the field alone, the legendary Galápagos finch researcher David Lack wrote half a century later. "However much information you can glean on the spot—about their ecology, their available food supply, their habits, the size of the different populations—population genetics if you like—breeding birds in cages and all the rest. You've just got to come back to the Rothschild collection and get down to measuring beaks." In other writings, Lack expressed an even greater debt to the CAS collection.[72]

The turn-of-the-century Galápagos explorers offered little theoretical analysis about the specimens they collected. The two key expedition leaders, Harris and Beck, never published anything about their work beyond field notes and travel journals. The Stanford students and CAS assistants contributed a few scientific articles about their Galápagos specimens but largely deferred to senior scientists at their respective institutions. These scientists and Rothschild did generate voluminous reports on the material brought back to them, but most of these simply identify the specimens and describe new species and varieties. They conspicuously avoided using the vast amount of new Galápagos material to tackle the pressing scientific problems surrounding the causes and mechanisms of organic evolution. "If such collections are not sufficient to throw light upon these problems," Rothschild asserted at the time, "it is not want of material that prevents our coming to satisfactory conclusions." Yet even he held back, content at his best with simply describing the pattern of evolutionary relationships linking Galápagos and mainland types.[73]

None of these scientific publications raised any doubts about the fact of evolution on the Galápagos. To the extent that they discussed the origin of Galápagos species, they uniformly found an evolutionary distribution pattern, with a large number of endemic species clearly related to Central and South American types. Most of these researchers reaffirmed the view of "Darwin, Wallace, and most other naturalists, that the islands were uplifted from the ocean, and never were in connection with the continent of America, or with each other."[74] A few chance migrants—birds, reptiles, plants, but no large land mammals—reached the rising archipelago over the centuries and evolved to fill the niches of life there.

In a joint paper, Snodgrass and Heller labored to reconstruct the family tree of Galápagos finches based on an apparent tendency toward blackness in the genus and extreme variability in beak thickness. "Hence, by plotting with the color variation as the ordinate and the bill variation as the abscissa, we can arrive at an approximate scheme of the relationships of the different subgenera, species and subspecies of the genus." With this, they theorized about which current species "stand nearer to the ancestral *Geospiza*" that first migrated to the Galápagos and how the genus spread over the islands. Yet they never speculated about whether isolation, the environment or natural selection caused these changes. Thus they ducked the central biological issue of the day, presumably because they had nothing to add.[75]

A similar approach marked a Tring Museum analysis of the evolutionary development of Harris's flightless cormorant from its flying ancestors. The published article feigned no hypothesis as to the cause of this evolution except to raise "the curious but rather suggestive coincidence" that many types of flightless birds "occur in seas of the neotropical region." The article then undermined this environmental explanation for flightlessness by adding that several such species lived near Antarctica, including "the little Duck *Nesonetta*, which, although not flightless, is apparently doing its best to reduce its wings to that condition."[76] This article reflected Rothschild's view generally. He clearly saw evolution in action throughout nature, but sentimentally clung to neo-Lamarckian explanations for it without marshaling sufficient evidence to reach a scientific conclusion.

To the extent that any scientific debate enlivened Galápagos studies during this period, it was over Baur's subsidence theory for the archipelago's origins. This issue usually arose in the context of trying to account for the source of Galápagos Tortoises: transoceanic migration could satisfactorily explain the presence or absence of most other species on the islands. Having occasionally seen giant tortoises survive in salt water, the explorers themselves tended to accept the majority view that these land reptiles or their eggs somehow floated or rafted to the archipelago. For example, Heller, who was committed to an oceanic, uplift explanation for the islands based on his study of Galápagos birds, hypothesized that shifting sea currents carried tortoises between their island homes in the Pacific and Indian Oceans. "Any land connection between these remote island groups," he asserted, "is not to be seriously considered."[77]

From his desk in San Francisco, senior CAS herpetologist John van Denburgh could not believe that these giants, which "are absolutely helpless in water," could have drifted to the Galápagos "at the mercy of winds and currents." Rejecting an oceanic origin for the archipelago on this evidence alone, van Denburgh concluded, "We must rather adopt the view that the islands are but the remains of a larger landmass which formerly occupied this region, and was inhabited by tortoises, probably of but one race." A faithful student of David Starr Jordan, van Denburgh attributed the "differentiation of species on the Galápagos Archipelago" solely to their isolation on separate islands. He never invoked natural selection or environmental factors. "Variation through a long period of time produced specific and sub-specific changes in these isolated colonies of reptiles, until each island . . . sustained its own peculiar kind."[78]

If the wealth of new material from the Galápagos could not finally resolve even the relatively straightforward scientific question of the islands' origins, then it appeared wholly inconclusive on the Gordian knot of the cause of evolution. Thus, these publications passed over this great question of turn-of-the-century biology. F. X. Williams, a Stanford student charged with collecting insects on the CAS expedition, speculated about the options in his unpublished notes. "How came the grasshoppers of the Galápagos to be differentiated as species + races?" he asked. "Either by *isolation* or *environment* or by *nat selection*." He then downplayed isolation because the separate species did not neatly segregate onto different islands, and questioned the environment because the entire archipelago seemed so similar to him. He dismissed the final alternative out of hand: "Natural selection I don't think plays an important part here."[79] No options remained. The flood of new material from the Galápagos reconfirmed the fact of evolution for researchers but left them wondering how it happened. Yet more than ever, the Galápagos Islands appeared to hold the answer. "If we are not able now to solve some of the problems alluded to," Rothschild wrote, "no accumulation of zoological specimens will ever help to answer our questions."[80] The issue demanded closer study.

SIX

THE AMATEURS
TAKE OVER

THE CALIFORNIA ACADEMY OF SCIENCES expedition of 1905–1906 marked the end of an era in Galápagos science. Its explorers, and those dispatched by Stanford University and Walter Rothschild, effectively completed the process of inventorying the archipelago for science, at least in gross terms. As a result of their efforts, Western scientific institutions possessed more comprehensive natural history collections from the Galápagos than from nearly any other place in the Pacific. Subsequent collecting expeditions to the region naturally focused on other island groups. The First World War disrupted international exploring during much of this period anyway. Further, it took two decades for naturalists to process the tens of thousands of specimens brought back by the turn-of-the-century expeditions to the Galápagos.[1]

More than anything else, however, a sea change in the natural sciences left the voyage of the *Academy* as the last great scientific expedition to the Galápagos. By 1900, laboratory experiments had displaced fieldwork as the preferred means to study life, with researchers increasingly calling their discipline *biology* and themselves *biologists*. Genetics, embryology, cytology, microbiology and the like came into vogue among scientists, and the best work in these departments was confined to research universities in the United States and Europe. "Naturalists" of the order of Cuvier, Agassiz and even Darwin no longer defined the cutting edge, and their fieldwork was derided as akin to stamp collecting. Fast-breeding laboratory fruit flies displaced Galápagos finches as the favored source of information about evolution. Field studies were left largely to museum curators and amateur naturalists, who were expelled to the periphery of the scientific enterprise.[2] The Galápagos Islands were cast out with them.

More than any other American of his generation, Henry Fairfield Osborn personified the continuing naturalist tradition within science. Fittingly, the two institutions most closely associated with him—New York's American Museum of Natural History and the New York Zoological Society—assumed the lead in Galápagos science during the 1920s and 1930s. The scion of two prominent mercantile dynasties, Osborn opted to take over vertebrate paleontology rather than the family businesses, and he largely succeeded. Just as his father had leveraged his social connections to build a railroad empire, Osborn did so to dominate the New York science establishment during the first third of the twentieth century. He became the founding dean of the Faculty of Pure Science at Columbia University and served as president of both the American Museum, which he expanded into the largest science museum in the country, and the Zoological Society, where he presided over the opening of the acclaimed Bronx Zoo. For the museum alone, he raised from his wealthy friends an unprecedented $11,000,000 for buildings and twice that amount for expeditions and exhibits, including several involving the Galápagos.[3]

Educated at Princeton University under the personal tutelage of the brilliant natural theologian James McCosh, Osborn saw purpose throughout nature. More than the evangelical McCosh or his own conservative Presbyterian parents, however, Osborn tended to identify God with nature, which he once described as "the visible expression of the divine order of things."[4] Evolution thus represented the manifestation of the divine and Osborn's quest to understand it underlay his interest in the fossil record. Beginning as a conventional neo-Lamarckian who believed that living things evolved by their own effort in response to their environment, he gradually forged his own theory of orthogenesis (which he called *aristogenesis*), which saw nature itself as a creative force endowing species with the hereditary potential to progress in predicable linear directions. He loudly denounced both religious fundamentalists who denied evolution and scientific materialists who attributed it to the natural selection of random, inborn variations in individual organisms. Osborn associated the latter view with experimental geneticists such as Columbia's trailblazing Thomas Hunt Morgan, who studied evolution by breeding fruit flies in a laboratory. For Osborn, nature had invigorating, near-spiritual qualities and it could only be understood on its own terms through field study. "Every day during my forty-eight years' observation," Osborn wrote late in his life, "I become more of a *naturalist*, less of a scientist, still less of a rationalist."[5]

Osborn's science, religion and sense of noblesse oblige combined to make him a passionate advocate of natural history expeditions. Like many Americans and Britons of his class and generation, Osborn feared that urbanization and industrialization were sapping the Anglo-American "race" of its native vitality. Redemption lay in confronting nature. The spirit of this age gave birth to the Boy Scouts and YMCA, movements to conserve natural resources and create national parks and the sports of hunting and fishing, in which the pursuer gave the pursued a fighting chance and often mounted or released the prey after capture. For Osborn, the same impulse found fulfillment in expeditions. "Except for a few early statistics," Osborn wrote about himself, "my biography actually begins with the first call to biology and geology."[6] Under his leadership, the American Museum and the New York Zoological Society capitalized on the enthusiasm and anxiety then associated with exploring and studying nature to send generously funded expeditions to exotic places.[7]

For Osborn and countless other Americans, Theodore Roosevelt exemplified the virtues of a vigorous life engaged with nature. Roosevelt hailed from the same circle of Gotham society as Osborn. Indeed, Roosevelt and Osborn's younger brother, Frederick, were close boyhood friends. As teenagers, they collected natural history specimens together—particularly by shooting birds—and frequented the nascent American Museum, an institution both their fathers generously supported. Roosevelt so loved nature that he considered specializing in science at college but was discouraged from doing so by the experimental turn that biology had taken. His teachers at Harvard feared that his poor eyesight would hinder microscope research. "They overlooked the fact that besides primordial slime and determinant chromosomes there were also in the world grizzly bears, tigers, elephants and trout," the naturalist David Starr Jordan later lamented, "all of which yield profound interest and are alike worthy of study."[8] Roosevelt retained a keen interest in game animals throughout his life, which was punctuated with a series of highly publicized domestic hunting and fishing trips during his presidency and two sensational foreign expeditions following his retirement in 1909—the first to hunt big game in East Africa for the Smithsonian, and the second to chart the upper reaches of the Amazon for the American Museum.[9]

Roosevelt's grand expeditions grabbed the nation's attention and set an example that others rushed to emulate. "Unless we may exempt his Conservation Policies, Roosevelt's greatest service during his presidency was the inspiration he gave young men," wrote Gifford Pinchot, the blue-blooded conservationist who managed the nation's forests during the Roosevelt administration. "To the boys of America he was all they hoped to be—a hunter, a rider, a sportsman, eager for the tang of danger, keen and confident, and utterly unafraid. There was no part of his example but was good for boys to follow."[10] Roosevelt himself aggrandized his exploits in popular magazine articles and best-selling books. For example, from the Sudan he wrote, "'I speak of Africa and golden joys'; the joy of wandering through lonely lands; the joy of hunting the mighty and terrible lords of the wilderness. . . . These things can be told. But there are no words that can tell the hidden spirit of the wilderness, that can reveal its mystery, its melancholy, and its charm."[11]

Theodore Roosevelt never visited the Galápagos Islands, but several of the amateur naturalists inspired by his example wore a path through the archipelago in the years between the two world wars. In 1929, for example, Pinchot topped off his public service as a Roosevelt conservationist and Pennsylvania governor with a glorious expedition to the Galápagos on behalf of the Smithsonian Institution.[12] A year later, Roosevelt's adventure-loving son Kermit led a collecting expedition to the islands for Osborn's museum. Even cousin Franklin got into the act, cruising through the archipelago in 1938 aboard an American warship during an official tour of Latin America. The wheelchair-bound president never disembarked on the islands, but he displayed his Rooseveltian vigor for the press corps by catching game fish for the Smithsonian collection.[13] A dozen other such "expeditions" passed through the Galápagos during the intrawar years, each striving to outdo its predecessors in social style if not in scientific substance. These expeditions captured the spirit of a sensation-loving age, and their explorers embraced Theodore Roosevelt's motto, "Life is a great adventure and the worst of all fears is the fear of living."[14] Properly packaged, a voyage to the Galápagos—all these were still sea voyages—could engender this highly romanticized sense of living.

A 1923 New York Zoological Society expedition set the standard for those that followed. Although sailing under the self-proclaimed motto "It's all for Science," this enterprise differed from any scientific expedition that preceded it to the archipelago. As leader of both the Zoological Society and the American Museum, Osborn saw the trip as a means to obtain Galápagos penguins, flightless cormorants and other exotic live birds for the former institution and stuffed specimens for two iguana exhibits at the latter. As time permitted, the exploring party also casually collected other natural history specimens. Photographers, artists and motion-picture camera crews went along to record the sights for zoo and museum displays. Time did not permit any original research in the archipelago, however, nor did the party's makeup.[15]

Zoological Society officials arranged the expedition on short notice at the whim of Harrison Williams, a wealthy patron.[16] During the twenties, Williams transformed a modest fortune based on manufacturing into a vast financial empire built on the electrical power boom. Shares of his Central States Electric Corporation split sixtyfold during the decade, pushing his personal net worth toward the billion-dollar mark before the stock market crash of 1929. Originally from Ohio, Williams bought a palatial mansion near the American Museum in New York and the world's largest yacht, the *Warrior*. "The only reason the Harrison Williamses don't live like princes," a *New York Times* article quipped, "is that princes can't afford to live like the Harrison Williamses."[17] He took up foreign travel and natural history as pastimes and became a major benefactor of the Zoological Society. In 1923, Williams offered to pay all the expenses for a collecting expedition to the Galápagos Islands led by William Beebe, the society's curator of birds and most famous personality.

Beebe took charge immediately, as he always did, organizing a lavish expedition for Williams and others aboard the 250-foot chartered yacht *Noma*. Taking advantage of the Panama Canal, which had opened in 1914, and *Noma*'s 2,500-horsepower engines, the so-called Williams Galápagos Expedition departed from New York harbor on March 1, 1923, and re-

turned just eleven weeks later. "This was just in time to rush the collections of live mammals, birds and reptiles to the Zoological Park, and to
frame and hang for exhibition the one hundred and thirty oil paintings
and water colors made during the trip, in readiness for the Annual Garden
Party of the Zoological Society on May 17," Beebe noted.[18] The *Noma*
spent only four days in the Galápagos, but this tight schedule was no
handicap. "The splendid group of young naturalists under the leadership
of William Beebe knew what facts to look for and where each fact belongs
in the still unfilled archives of present and past Evolution," Osborn
boasted. "Thus among the scientific wonders of our century, we have to
record: first, that in less than one hundred actual hours on land, the *Noma*
party accomplished results—artistic, photographic, observational—which
are entirely without rival."[19]

Osborn here used the term *naturalist* loosely. Except for the Harvard
entomologist William Morton Williams, who briefly joined the expedition in Panama, Beebe was the only person aboard the *Noma* who
answered to that description, and even he lacked so much as an under-
graduate degree in science. In 1899, Osborn had plucked Beebe from
among his younger students at Columbia University to work at the new
Bronx Zoo; it was an inspired choice. Beebe developed a rare knack for
communicating science to the public, beginning with exhibits and lectures
but eventually graduating to articles and books. By the time of the
Williams expedition, Beebe had explored four continents, published eight
books and become a regular contributor on science to the *Atlantic Monthly*
and *Harper's*.[20] Professional zoologists withheld their favor by denying
Beebe election to the National Academy of Sciences despite Osborn's
pleas, but the public accepted him as a voice of science. He easily assembled volunteers for his expedition to the Galápagos. Besides the paid crew
and Zoological Society personnel, the *Noma* carried a half dozen New
York business and social leaders, including the first women to accompany
a scientific voyage to the Galápagos since Elizabeth Agassiz—the artist
Isabel Cooper and the former actress Ruth Rose. "It was a wonderful trip
from the first," Rose told reporters. "It was just the greatest sport in the
world!" A highlight for her came in helping a former assistant secretary
of the treasury catch a fish in his hat. "It was all for Science," she said
laughing.[21]

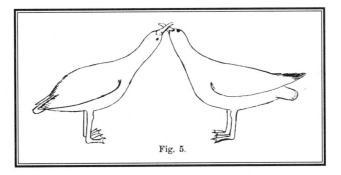

Fig. 5.

Sketch of waved albatross mating dance, from Frederick
Drowne's diary in Novitates Zoologicae *(1899)*
(Source: Used by permission of UGA.)

The Williams expedition provided the fodder for Beebe to write ten articles and the book *Galápagos: World's End.*[22] "His popular titles were highly readable and sold well," Beebe's biographer notes, "even though there was an element of fiction in some of the nonscientific material they contained."[23] This element, which was present in the scientific content as well, contributed to their appeal. In writing about the Williams expedition, Beebe lavishly embellished the narrative and grossly exaggerated the trip's scientific significance, for instance in his account of an excursion to Albemarle Island's Tagus Cove that netted two flightless cormorants for the Bronx Zoo.

This particular outing left from the *Noma* promptly after breakfast, and had to return by ten. "It was too chilly for comfort," the intrepid naturalist complained. Most of the party went by dinghy but Beebe and two others climbed overland "up and down, up and down, over slopes just under the sliding point of loose clinkers." To complicate matters, Beebe carried a 40-pound movie camera, a shotgun and "a game-bag which was diabolically clever in getting in wrong places." Several times he slipped "with a movement as sickening as that of a circular earthquake. Rock after rock would hurtle to the bottom and splash into the black, sharkful water." The explorers had set out that morning to film and catch birds at a rookery where they hoped to find the reclusive cormorant, described in *Galápagos: World's End* as "a bird probably doomed in a few years to an extinction as complete as the great auk's." Upon spying a nesting female, "I seized her as quick as possible," Beebe related, "but quick as a flash she dodged, leaping off her nest and when I caught her body, she raked my hand fore and aft with the cruel curved tip of her beak." A colleague "rescued the single egg" for science. Beebe erroneously reported this as the first live capture of a species so elusive as to have "escaped the attention of Darwin." Ruth Rose, who actually spotted the bird first, described the excursion as "almost too much joy."[24] Cormorants became the shore-hugging expedition's prize zoo trophy when its sole tortoise died after Beebe tried to test Baur's subsidence theory by seeing how far the animal could swim in seawater.[25]

Beyond stirring public interest in natural history expeditions, the Williams expedition contributed little to science; it brought back the requisite museum and zoo specimens, numerous pictures and films of Galápagos animals and a small research collection composed mostly of insects and fish. Zoological Society scientists identified only a few dozen new species from this collection, but proudly announced each one in their in-house research journal.[26]

Beebe's greatest service to science was as a popularizer, and in this he reigned supreme. "If modern science is determined to leave no mysteries," *The Nation* editorialized, "let William Beebe tear away this last veil, for there is no man better equipped to discover the glamour of reality when illusion is gone."[27] Beebe's published accounts of the Williams expedition espoused evolutionism of a near religious variety. In this he followed Osborn, whom Beebe hailed in *Galápagos: World's End* as "the high priest of some uncomparable religion." Like Osborn, Beebe favored neo-Lamarckian and orthogenetic explanations for the origin of species. He suggested, for instance, that the native cormorant, secure in its Galápagos abode, lost the power of flight by not using it and that the marine iguana evolved from its terrestrial ancestor along fixed lines. As for the different species of finches that he found feeding together on the islands, Beebe concluded that variation in their beaks "is not directly adaptive, but reflects a relaxed environmental control." Variation alone apparently caused their speciation.[28] Behind these words lay muddled assumptions about the cause of evolution: for the cormorants, individual effort; for the iguanas, a kind of predetermination; for the finches, aimless drift.

Beebe thus muted the struggle for survival and presented nature as a benign womb for the human species. "I realize more than ever what a casual thing is man upon the earth," he reflected in *Galápagos: World's End.* "Once we were taught that the earth was the center of the universe; then that man was the raison d'être of earthly evolution. Now I was thankful to realize that I was here at all, and that I had the great honour of being one with all about me, and in however small a way to have at least an understanding part."[29] In his happiest discovery of the Williams expedition, Beebe claimed to find that by some "environmental insular relaxation" the absence of predators on the Galápagos naturally led native species to produce fewer offspring than their mainland relatives, thereby averting a Malthusian struggle among siblings and clearing the way for a gentler evolution than that envisioned by Darwin.[30]

Fortuitously for Beebe, the Williams expedition occurred just as the American controversy over creation and evolution reignited. This fed popular interest in the cruise. By 1923, persistent religious objections to Darwinism had crystalized into a nationwide crusade, led by fundamentalist orator and politician William Jennings Bryan, against the teaching of human evolution in public schools. Osborn took the point position in countering Bryan through lectures, articles and books.[31] Beebe fell in line, studding his Galápagos narratives with barbs against "Bryan and his anthropocentric isolation, the fear and the egotism of his aloofness from our lesser blood brothers." The media loved it, and so did the expedition's sponsors.[32]

No sooner had the *Noma* docked in New York than the Zoological Society began planning a longer expedition to the archipelago for 1925. Beebe again assumed command, this time joined by his Madison Avenue publisher, George Putnam, who brought his family. Delighted with the fulsome credit accorded him for the 1923 expedition, Williams offered to underwrite the follow-up voyage. Other noted philanthropists lined up to contribute as well, including Vincent Astor, Clarence Dillon, J. Pierpont Morgan's son Junius and Marshall Field. Industrialist Henry Winton lent his fabulous 2,400-ton steel yacht *Arcturus*, which the Zoological Society refit for deep-sea exploration. "It is equipped for dredging and trawling," one magazine reported, "with a sort of cow-catcher in front to facilitate operations with the net and the harpoon-gun."[33]

With the *Arcturus* at his disposal, Beebe's second expedition to the Galápagos spent more of its time on the open ocean than within the archipelago and again made only brief visits ashore. Conditions were simply too ideal on board for members of the expedition to want off. In addition to seventeen explorers and passengers, the *Arcturus* sailed with a crew of thirty-five, including chefs, stewards and two wireless operators assigned to send daily press reports to the *New York Times*. Newspapers around the world followed the expedition's progress, often giving it front-page coverage. When not sending messages, the ship's massive communications equipment tuned in pioneering commercial radio broadcasts from the United States. Beebe's party even radioed song requests from the islands to WCBS in New York. Beebe still billed the Galápagos as the "world's end," but this was a far smaller world than ever before. "The radio was an important and popular part of the ship's equipment," one of the operators noted. "When we were not sending press dispatches or tending to the business of the expedition we were kept busy with the love-and-kisses type of message."[34]

Beebe regularly dispatched extended articles about the expedition to the *New York Times* for publication in its Sunday magazine. These self-aggrandizing pieces became the basis for another book, *The Arcturus Adventure*. In one of these articles, "The Ocean Tells New Tales to Beebe," the naturalist described his use of a diving helmet for underwater research. "As I peer out through my little rectangular windows I seem to be actually living an experience which only the genius of a Verne or a Wells can imagine into words," he wrote. "I have even the sensations of a god." Beebe's claim of being the first person to use such apparatus for scientific research brought a public rebuke from an indignant colleague, who cited a six-year-old American Museum publication describing the technique. Yet Beebe's descent was the first use of diving equipment in Galápagos waters and inspired others to experiment with the procedure elsewhere.[35] In another Sunday magazine article, "Beebe Climbs a Fiery Volcano," Beebe related his thrill at witnessing a volcanic eruption from aboard the *Arcturus* and his brush with death when seeking a closer look. "Baked from above and below, we staggered on," he wrote of his climb toward one crater, "unable to sit down and rest for the intolerable heat of the rocks." His published claim that "this was the first eruption known to men" on the Galápagos ignored earlier reports, including one by a maternal relative of Zoological Society patron Franklin Delano Roosevelt. The story became front-page news around the world, however, and added to the growing Galápagos mystique.[36]

Once again Beebe brought back little of value for serious researchers. Yet his lanky frame amid Galápagos iguanas became the image of science for many Americans. "There is a kind of research which consists in breeding countless generations of fruit flies," the *New York Times* review of *Arcturus Adventure* asserted. "This sort of science is about as comprehensible to the man in the street as is the philosophy of a flagellant monk. But the man in the street can understand William Beebe, because Mr. Beebe is a wanderer and an adventurer, and we are all homesick for wanderings and adventures." Theodore Roosevelt had made similar comments in his *New York Times* review of an earlier Beebe book: "It will stand on the shelves of cultivated people," the former president wrote, "as long as men and women appreciate charm of form in the writings of men who also combine love of daring adventure with the power to observe and vividly to record the things of strange interest which they have seen."[37]

Theodore Roosevelt had been Beebe's idol and Osborn his mentor. They anchored a vibrant circle of New York naturalists: adventurers who defied the current running toward laboratory research in biology. When others criticized Beebe for his errors and excesses, Osborn defended him. "We cannot work a star of the first magnitude as we would a cart-horse," he once advised a fellow Zoological Society officer. "We must realize that we have in Beebe a star and the makings of one of the greatest naturalists of our time." Experimental biologists might disagree with this assessment, but Beebe's popular accounts of the Galápagos Islands launched countless ships to that remote archipelago and for a generation were more widely read than those written by Darwin.[38]

So many private yachts sailed for the Galápagos during the ensuing decade that their paths often crossed in the supposedly desolate archipelago. Adventure-seeking travelers also began encountering a new breed of settler on the islands, as a few hardy Europeans sought refuge there during the lost years between the two world wars. Most of the yachting crowd went simply to sightsee or fish, but some of them emulated the Williams expedition by hosting a veneer of naturalists, having a zoo or museum sponsor and publishing accounts of the adventure. As an added incentive for wealthy Americans, sponsorship by a scientific or educational institution made the entire cost of the expedition deductible as a charitable contribution under the nation's new income tax code.[39]

By 1930, four more such expeditions reached the Galápagos Islands from the United States, all of them using Beebe's books as their guide. On his annual cruise for 1926, philanthropist William Vanderbilt passed through the region aboard his yacht *Aca* with a party of friends and naturalists gathering fish for his immense private collection and land animals for the American Museum. "Have visited Tagus Cove and find same up to expectations as you described it in your book," he radioed to Beebe about the cormorant rookery.[40] Two similar expeditions to the South Pacific collected museum specimens in the Galápagos three years later, the first aboard a three-masted schooner captained by Gifford Pinchot and the second led by Cornelius Crane, the twenty-three-year-old grandson of Chicago industrialist R. T. Crane, sailing the brigantine yacht *Illyria*. In the book *To the South Seas*, Pinchot described his voyage as "adventure seasoned with science." The task of memorializing the Crane expedition in literature fell to the leader's Harvard College classmate and traveling companion Sidney Shurcliff, who wrote in *Jungle Islands*, "At first Cornelius had intended to take a group of friends on purely a yachting cruise. But as the *Illyria* was designed with ample accommodations, and as her owner was notably fond of hunting and fishing, it was decided that various experts in zoology, natural history and the like should be included in the party." Crane's collection was earmarked for Chicago's Field Museum while Pinchot's went to the Smithsonian. Neither amounted to much.[41]

The Astor expedition of 1930 surpassed the other three in quantity and quality of its collections. "This trip, organized by Vincent Astor, was in a class by itself for efficiency, comfort, and luxury," one participant wrote. "We were fortunate enough to have the finest private ship afloat, . . . the *Nourmachel*."[42] Astor, who then controlled one of America's greatest family fortunes, equipped his yacht with electrically heated tanks to transport live Galápagos fish to the New York Aquarium and afforded an American Museum exploring party ample time to collect tortoises in the remote interior of Indefatigable Island. "The return from the Galápagos Islands of another expedition of scientists and pleasure-seekers . . . shows that the spell of the archipelago is still potent," the *New York Times* commented in 1930. "This last of a half a dozen expeditions from New York in ten years

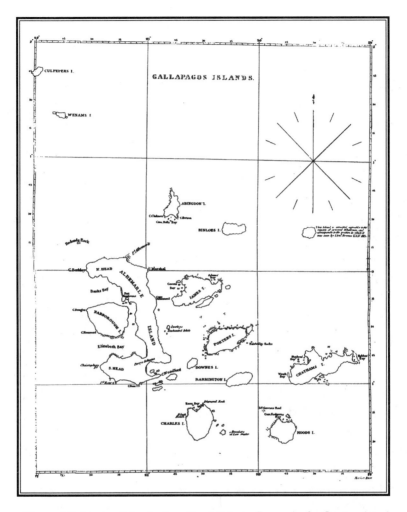

Chart of Galápagos Islands, from David Porter, Journal of a Cruise *(1822)*
(Source: Used by permission of UGA.)

will bring back some treasures, among them great land tortoises for conservation, tanks of coral reef fish, a Pacific green turtle, four young sea lions and a marine iguana."[43]

Beebe did not launch this parade of ships; he simply gave it a destination. The multiplicity of these expeditions resulted from the yachting craze that swept East Coast high society following the First World War. Wartime advances in shipbuilding technology contributed to this, as did a

postwar surplus of fast, light warships—some of which became private
yachts. Wealthy Easterners vied to own ever larger and more luxurious
ships, and needed a place to go with them. Beebe's popular writing did the
rest, coupled with the accessability of the Galápagos from New York
through the Panama Canal. "From gayest Monte Carlo to wildest Galápagos, Yankee yachting palaces lie at anchor," a society columnist wrote in
1930. "It is not so much that the wealthy landlubber has suddenly taken to
the sea as that he has conquered it. Between the upholstered, interior-
decorated salons of his home afloat and his Park Avenue apartment
ashore, there is little visible distinction."[44]

Even sailing in the wake of Beebe, many yachters went to the Galápagos Islands without any pretensions of doing science. Hundreds of people
on dozens of private vessels and at least one regular cruise ship visited the
place each year during the late 1920s. Some simply wanted to see strange
animals and fiery volcanos; others went primarily to fish. Deep-sea rod
and reel fishing had come into vogue along with yachting.[45] Beebe included an entire chapter on deep-sea fishing in Galápagos: World's End,
which immediately caught the attention of America's chief chronicler of
sport fishing, Zane Gray. Although he gained fame and wealth writing
Western fiction, Gray always preferred to write about his own fishing trips
and eventually published over a hundred articles and seven books about
them.[46] He set his sights upon the Galápagos after reading Beebe's account and was there within a few months. "Strange new world this Galápagos Archipelago," he wrote in a 1925 book about his trip. "I had sailed
four thousand miles to fish virgin seas that embraced these islands. Sight
alone of the volcanic slopes and coves, and the purple channels that circle
these shores, would have been ample reward for the long journey."[47]

Although fishing in the Galápagos did not fulfill his highest hopes,
Gray did catch near-record tuna, wahoo, mackerel, grouper and shark, and
his writing about it further fueled popular interest in the archipelago.[48]
More American anglers went to the Galápagos during the late twenties
and early thirties than ever before. When he followed this crowd, philanthropist George Vanderbilt twice took along ichthyologists collecting fish
specimens for Philadelphia's renowned natural history museum. In 1929, a
group of wealthy American sport anglers tried to buy or lease the archipelago from the cash-strapped Ecuadorian government as a fishing and

wildlife preserve. Coming after several highly controversial attempts by the United States government to secure the Galápagos as part of the western defenses for the Panama Canal, the offer only served to stiffen Ecuadorian resolve to assert dominion over the islands.[49]

To counterbalance the "Yankee invasion," the government in Quito was already courting European investment in the Galápagos.[50] The most notable such development occurred in the late 1920s when a group of over 100 Norwegians, inspired by the romanticized accounts in *Galápagos: World's End*, attempted to colonize Charles, Chatham and Indefatigable Islands, which they called by their Ecuadorian names of Floreana, San Cristòbal and Santa Cruz. These colonies quickly foundered in the harsh realities of the Galápagos environment, leaving behind only a few stubborn Norwegians and a rusting cannery.[51]

The growing Galápagos mystique also lured a counterculture element from Europe. Two such parties on Floreana Island—a "Nietzschean" German dentist with his patient-turned-lover and a "neurotic" Austrian baroness with her two German lovers—gained international tabloid fame during the early 1930s as "Adam and Eve of the Galápagos" and "the mad empress and her court." Press reports described both parties as nudists, which made the cactus-strewn Floreana a particularly odd choice, but they seemed happy there. The dentist named his abode "Eden" and the baroness called hers "hacienda 'Paradise.'" A steady stream of yachters reported on their exploits until a sensational series of mysterious tragedies ended it all, less than five years after it had begun. In 1934, they all died or disappeared in rapid succession except for the dentist's distraught lover. A few less-eccentric Europeans lived on the same island during the thirties: families eking a "back-to-nature existence" out of the rough terrain. Larger groups of Central and Eastern Europeans sought permission from Ecuador to settle there as well, but never carried out their plans.[52]

Although no one kept precise records, the number of yachting expeditions to the Galápagos clearly declined as the economic impact of the Great Depression spread worldwide. New Yorkers in particular largely lost interest in the place during the thirties. Those who carried on—mostly from California—raised the standard of scientific research, however, and at the decade's end a lone Englishman put the Galápagos back in the center stage of science.

As wealthy New Yorkers faded from the scene, California entrepreneur G. Allan Hancock picked up the slack. After migrating to San Francisco during the 1849 Gold Rush, Hancock's father foresaw greater economic opportunity in Southern California. Near the growing town of Los Angeles he bought an enormous ranch that happened to include the La Brea tar pits. In addition to raising crops, the Hancocks sold tar from the pits for roofing and paving. Allan Hancock gradually took over the family businesses from his mother after his father died in 1883. With the coming of automobiles to Los Angeles, Hancock tried drilling an oil well at the tar pits. It came in a gusher.

Soon Hancock branched into banking, construction, transportation and other enterprises that boomed with the local economy. After depleting his oil deposits, he subdivided his ranch into residential and commercial parcels that became Hollywood and the upscale Wilshire district of Los Angeles. Rich beyond his wildest dreams, Hancock supported a diverse array of California educational and cultural institutions. He donated the land for the University of Southern California campus, for example, and played cello in the Los Angeles Symphony Orchestra. The discovery of well-preserved Ice Age remains at his La Brea tar pits around 1900 turned Hancock's attention toward natural history, and he began collecting fossils along the Mexican coast from his yachts *Velero I* and *Velero II*.[53]

Hancock first visited the Galápagos Islands in 1927. He had acquired a war-surplus Victory ship to transport fruit and vegetables from his Mexican ranches to California markets and earned a license to captain the commercial vessel himself. At the request of the California Academy of Sciences, "Captain" Hancock detoured through the Galápagos to gather specimens for a museum display. With Joseph Slevin of the 1905 CAS expedition as his guide and a well-timed volcanic eruption for entertainment, Hancock became enchanted with the place. When the fledgling San Diego Zoo then asked him to go back for live animals, he decided to construct a state-of-the-art, 200-foot-long cruiser, *Velero III*, specifically designed for scientific research in the eastern Pacific. On the ship's maiden expedition in 1932, Hancock and a team of naturalists collected Galápagos animals for the San Diego Zoo and the CAS's Steinhart Aquarium. After extensive dredging, trawling and research equipment were added to the ship's inventory, later expeditions specialized in studying shallow-water marine invertebrates.[54] When members of a small party collecting specimens for the Danish National Museum first saw *Velero III* slip into a Galápagos bay in 1934, they thought it was a navy destroyer. "The ship was a perfect miracle with its soft-carpeted salons, Steinway piano and luxurious single cabins with real beds and bathrooms," one of the envious Danes drooled. "Modern laboratories, huge aquariums and every up-to-date accessory were at the disposal of those investigating nature."[55]

The *Velero III* carried scientists on research trips to the Galápagos and eastern tropical Pacific for an average of three months each winter from 1934 until the United States Navy requisitioned the ship during the Second World War. Marine biologists from across America vied for positions on these luxury cruises during those otherwise lean years of the Depression, and often did their best work on board. Shipboard naturalists continued to collect live animals and preserved specimens for California zoos, aquariums and museums as well.[56] While Hancock captained the ship with its eighteen-person crew and tended his businesses back home via the radio, the scientists settled into a comfortable routine. "We breakfast in our work clothes at eight and when finished find the dredge boat, fish boat, and two or three skiffs in the water," one of them wrote. "At noon those without lunches are collected for lunch and if not finished with the location, return in the afternoon. If through, the afternoon is spent sorting and classifying the morning's collections." Scientists and passengers dressed in formal attire for dinner each evening. "We dine as one would at his club, at six, and after dinner we have one or two hours of classical music by the [shipboard] trio or quartette, followed by motion pictures." Hancock played cello in the after-dinner concerts and was known to choose scientists for his expeditions on the basis of their musical talents. It was good work for those who got it. In a parting tribute, one member of Hancock's 1934 expedition spoke of "the joys, the romance, the adventure of such a trip, with such a captain, such a mate and such a crew, on such a ship."[57]

In 1931, just as Hancock launched his series of annual trips aboard the *Velero III*, Templeton Crocker offered to take an expedition from the California Academy of Sciences anywhere it wanted to go on his sailing yacht *Zaca*. Crocker's grandfather, the San Francisco railroad baron Charles Crocker, had created the CAS's endowment fund. His father had presided over the academy's board of trustees for years prior to his death in 1897, and his cousin had filled that post ever since. Having recently returned from a pleasure cruise around the world aboard the *Zaca*, Templeton Crocker felt it was his turn to do something for science.

The timing of Crocker's offer was ideal for Galápagos science. After a quarter century of delays caused by CAS ornithologists' complete inability to make any sense of the near 5,000 specimens of Galápagos land birds brought back by the 1905–1906 expedition, the academy's new curator of birds, Harry S. Swarth, finally published a comprehensive report on the entire collection at almost precisely the same time as Crocker made his offer. This 300-page monograph focused on the islands' puzzling finches, which Swarth made all the more perplexing by his analysis. A consummate splitter when classifying birds, Swarth divided Galápagos finches into forty distinct species and subspecies, arrayed among five genera within one new family. Their beaks caused most of the trouble. He measured thousands of specimens, first at the CAS and then at Stanford, Tring and the British Museum. "Such remarkable extremes of variation in bill structure as are seen," Swarth wrote, "lie outside my experience with any North American mainland bird."[58]

A century earlier, seeing such diversity among Galápagos finches had inspired Darwin's thoughts about evolution—but when he could not trace his specimens to separate islands, he turned to Galápagos mockingbirds (which more neatly divide into island-specific types) as better evidence for his theory. Later expeditions deepened the mystery: everyone reported finding slightly different types of ever so similar finches living together, yet evolution by natural selection suggested that only the fittest of these species should survive. The enormous CAS collection seemed to confirm earlier reports. "In other words, natural selection was eliminated as a factor in the production of the observed variations," Swarth emphatically declared. "There are large bills and small bills, heavy bills and slender bills, among the ground feeding species of *Geospiza*, and also, pushed to nearly as great extremes, among the tree-frequenting genera." He could not resolve what caused the evolution of so many species from one ancestral type because neither environmental differences nor isolation operated as factors: perhaps it was simply extreme variability in the family, without sufficient competition to eliminate intermediate types. "The Galápagos Islands offer an unrivaled opportunity for further field work" on this problem, Swarth concluded. Periodic collecting of finch specimens from the same sites could help, he suggested, but even more "might be learned from carefully directed observations of the living birds amidst natural surroundings."[59]

Crocker's offer provided the CAS with an opportunity to monitor developments within the Galápagos finch population since 1905–1906, fill gaps in its general collections from the archipelago and obtain tropical fish for its Steinhart Aquarium. These three objectives sent the CAS back to the Galápagos for the first time in nearly thirty years. Along with Swarth, the academy sent botanist J. T. Howell and two fish experts. Crocker led the trip, of course, but he did not invite along any friends or family members. This was to be a scientific voyage, with the *Zaca* and its expert crew entirely at the disposal of the naturalists on board.[60]

"The yacht no longer has the appearance of a pleasure craft. Huge tanks have been installed on her decks to bring back live fish," a local newspaper reported on the day of the expedition's departure. "Below decks, the Zaca has been changed into a scientific laboratory." The ship still sailed in style, however, following "a bibulous send-off . . . lubricated by the 'Zaca cocktail,'" which Howell described as "a truly delicious and devastating concoction of gin and crushed fresh pineapple." Prohibition officials discreetly looked the other way. "My chief distress on the trip (and it was minor) was the occasional curtailment of my collection activities ashore," Howell later recalled. "Time and again my desire to climb to the highest point on an island was frustrated by Mr. Crocker's rule that everyone who went ashore had to be back on board by dinner time." Crocker made one exception to this rule when he joined the naturalists in making the first ever ascent to the summit of Indefatigable Island, which opened virgin territory for collecting. Except for that singular three-day excursion, Howell noted, the daily routine rarely varied. Crew members accompanied naturalists on their day trips and did most of the heavy lifting. "Dinner was never served until after dark," Howell added, "and it was preceded by a ritual of a dry martini and a capsule of quinine" to prevent malaria. "After dinner there was often phonograph music from a well-stocked musical library,—or if it had been a day afield, an hour or so of work before retiring." Pointing to the 400 stuffed or frozen birds, 3,000 plant specimens and 331 live fish brought back, CAS officials declared "this expedition would be of great

Sketch of Indefatigable Island, from Frederick Drowne's diary in Novitates Zoologicae *(1899)*
(Source: Used by permission of UGA.)

and permanent value to science." For his part, Crocker described it as "full
of adventures," and promptly offered to host a follow-up expedition that
would pass through the Galápagos in 1934 on its way to Polynesia.[61]

In part because it went to out-of-the-way sites to broaden the CAS collection, and because it happened to arrive during a period of unusually wet
weather, the Templeton Crocker Expedition obtained specimens of several
never-before or rarely collected species. Swarth nabbed a type of finch last
collected by Darwin, for example, and Howell found the typically arid
landscape green with vegetation. "It was 'hay-day' for botany!" Howell
cried.[62] "Before going to the Galápagos, I was told by a botanist who had
been there that there was nothing there for me to obtain, that everything
in that plant line had been done," he recalled. Yet the heavy rain spouted
dormant seeds and spurred many plants into full bloom: ideal conditions
for a field botanist who used flowering parts to classify plants. The Galápagos Islands were famous for their native cacti, for example, yet naturalists had identified only four different types, and some researchers
questioned whether even these four were truly distinct species. "Unusual
rains made the desert bloom like a garden and one of the advantages
reaped was [cacti] in abundant flower and fruit," Howell reported. "I have
come to recognize 7 species and 2 subspecies, representing 3 new species, 1
new specific name and 2 new subspecies. If nothing more had been obtained, these would have made the trip worthwhile!"[63]

Howell drew on evidence of these and other Galápagos plants to join
the debate over the archipelago's origins. On the basis of the even distribution across the islands of novel species clearly derived from American
types, Baur had proposed that the Galápagos were once part of a vast lost

continent linked to the mainland; preoccupied with tortoises, some earlier CAS researchers had agreed. Seeing the archipelago in full bloom, however, Howell could appreciate how a few introduced American plants could spread across the islands in lush times before evolving separately on each, and also that the Galápagos lacked too many basic types of American plants to have ever been linked to the mainland. At least to Howell's satisfaction, these observations confirmed Darwin's theory that the islands arose from the ocean floor—a view that regained ascendancy among scientists following the expeditions of the 1920s and 1930s.[64]

Seeing the Galápagos in fat times did not resolve the greater question of how those chance introductions evolved into the varied island types. Swarth returned more bewildered than ever about the finches, and Howell joined him in rejecting natural selection as the final answer. "The geneticist has shown us how plants may vary through mutation and hybridization," Howell observed. "But out on the Galápagos Islands, which I believe to be oceanic in origin, where probably Mollugo became established by the introduction of a single plant and where Alternanthera by one or two plants, I cannot look at the present array of Mollugos or Alternantheras without feeling the presence of a formative force in the environment, which, not only makes adaptations in the soma, but make an hereditary impression on them on the germ." Howell did not deny the primacy of inborn genetic variation in evolution generally, but his encounter with Galápagos flora persuaded him that the environment could cause—as well as select—hereditary changes.[65]

Howell was not alone in maintaining this view. At least into the 1940s, neo-Lamarckism survived as the favored explanation for evolution in Darwin's archipelago. During the 1930s and early 1940s, even such a prominent Darwinist as the British biologist and popular science writer Julian Huxley downplayed the role of natural selection there. "It would rather seem that the prevention of intercrossing between the populations of the various islands has left each stock free to develop along its own lines, and that each has happened upon a slightly different one," Huxley wrote about the Galápagos in 1931. "Oddities flourish in out-of-the-way places. The primarily conservative disposition of Natural Selection spares the corners."[66] Yet reproductive isolation could not account for the evolution of island-hopping finches.

Despite the growing sense that evolution might operate differently on remote islands than on continents, field naturalists continued to believe that Galápagos plants and animals held important clues to understanding the origin of species generally. Following the Templeton Crocker Expedition, for example, Howell told a scientific audience, "It would appear that there are few places in the world where [these] problems can be studied under such favorable and unusual conditions—a fact which has led me to call the archipelago Evolution's workshop and showcase."[67]

So many naturalists agreed with this view of the Galápagos that in the late 1920s they began taking steps to preserve the archipelago in its natural state and protect its endangered wildlife. The New York Zoological Society initiated an ambitious scheme to breed Galápagos Tortoises at various subtropical sites in the southern United States, and for that purpose took 181 of them off Albemarle Island aboard a U.S. Fisheries Bureau vessel in 1928.[68] Although the breeding colonies failed, some of the tortoises still survive in American zoos. Swarth returned from the Templeton Crocker Expedition urging that Ecuador set aside the islands as a wildlife sanctuary. The British Museum ornithologist P. R. Lowe took up the cause in England, arguing that Galápagos finches presented "a biological problem of first class importance, and that this problem alone would justify the establishment of biological reserves on one or more of the islands."[69]

The Ecuadorian government responded in 1934 by outlawing the killing or capture of tortoises, fur seals, sea lions, iguanas and certain birds in the Galápagos and, two years later, by setting aside uninhabited portions of the archipelago as nature preserves.[70] Recognizing that the mainland government could not enforce even these limited restrictions in the remote islands, the itinerant American travel writer Victor von Hagen proposed establishing a permanent science institute in the archipelago to coordinate research and monitor compliance with conservation laws. "I personally visualize a station on Indefatigable, the center of the islands—an unpretentious institution that will work in cooperation with visiting expeditions," he wrote.[71] Huxley thought that at least a local game warden was needed for the Galápagos and worked with the wealthy and well-connected American conservationist Harold Coolidge to secure international support for one.[72] Neither proposal got past the planning stage for more than a quarter century.

Based on his reading of Swarth's monograph and his own work with the British Museum bird collection, Lowe joined Swarth in urging further field study on Galápagos finches. "They represent A HETEROGE-NEOUS SWARM whose diversity has been the despair of systematists," Lowe told the British Association for the Advancement of Science at its 1935 Darwin centenary celebration. Certainly different types of similar birds can coexist, he conceded, but nowhere do they live in the same place on identical food "no matter how far afield we roam in our search." This would fly in the face of natural selection—yet it was precisely what explorers claimed for Galápagos finches. Lowe's best guess blamed the "bewildering diversity, intergradation, and distribution" of the finches on rampant cross-species breeding that produced healthy hybrids "uncontrolled by selective action"—a highly unusual phenomenon in nature. "There is no group of birds in the whole world which has more right to occupy the attention of zoologists at the present moment," he asserted, and urged that "properly qualified investigators . . . be sent to the Galápagos with the sole object of studying on the spot and for a sufficiently long period, by means of *actual breeding experiments*, the genetics of this very interesting group of birds."[73] Huxley, then secretary of the Zoological Society of London, knew just the person for this job—David Lack.

A twenty-five-year-old English schoolteacher and amateur bird watcher at the time, Lack had come to Huxley's attention as a gifted observer of bird behavior. Huxley arranged for the Zoological and Royal Societies to pay Lack's expenses for a bare-bones expedition to observe finch behavior on the Galápagos over the course of one entire breeding season. He would test Lowe's hybridization theory.[74] Finally departing in 1938 after a long delay, Lack and five companions traveled by commercial steamers and stayed with local settlers. "The Galápagos are interesting, but scarcely a residential paradise," he later wrote. "The biological peculiarities are offset by an enervating climate, monotonous scenery, dense thorn scrub, cactus spines, loose sharp lava, food deficiencies, water shortage, black rats, fleas, jiggers, ants, mosquitoes, scorpions, Ecuadorean Indians of doubtful honesty, and dejected, disillusioned European settlers." Money was tight and food insufficient. One of Lack's companions almost died of dysentery. They never saw a yacht.[75] Despite his personal discomforts (or perhaps because of them), Lack did see something on the Galápagos that no one had ever seen before—natural selection at work among its finches through interspecies competition.

This discovery took time. Lack's party stayed at the settlement on Chatham Island for the first month, which "proved to be dismal"; then for three more months at Academy Bay on Indefatigable, where "no European settlers would speak to any others of different nationality." Lack spent his mornings observing the feeding and breeding habits of finches in the wild and his afternoons building aviaries in a failed attempt to get the estrous birds to mate across species. When the breeding season ended in 1939, the party caught thirty-one live finches from four different species for continued breeding experiments and headed back to England. "These [birds] traveled badly," Lack commented, so he detoured with them to San Francisco and left them in the care of the CAS. He stayed on for five months examining the CAS finch collection; then went to study Rothschild's finches at the American Museum in New York, where he roomed with the collection's curator, the German émigré zoologist Ernst Mayr, just as the Nazis began bombing London.[76] In total, Lack measured the dimensions of some 8,000 finch specimens, including the length, width and depth of their beaks.

At the time, Huxley and Mayr stood at the forefront of a select group of biologists forging the so-called neo-Darwinian synthesis. Both contributed influential books to the effort in 1942, and profoundly affected Lack's thinking about Galápagos finches. The synthesis brought together the work of laboratory geneticists and field naturalists. Geneticists introduced the conviction that evolution took place solely through the natural selection of genetically based variations in organisms. When stripped of neo-Lamarckian and religious notions of these variations occurring after the organism's conception or as a result of supernatural forces, the theory became known as neo-Darwinism. Naturalists added the view that a species is not a group of recognizably similar individuals but rather a reproductively isolated population filling a specific ecological niche. Together, these two perspectives refocused scientists' thinking about evolution. Now there was a fixed physical process, inborn genetic variation, on which natural selection acted; and the entity that was evolving—the species—was defined in terms of a breeding population rather than discrete organisms. This is the Modern Synthesis. "A new species develops if a population which has become geographically isolated from its parental species acquires during this period of isolation characters which promote or guarantee reproductive isolation when the external barriers break down," Mayr proposed in his 1942 treatise. During the darkest days of the world's worst war, the light of natural selection reemerged.[77]

Lack did not initially fit the Galápagos finches into the new synthesis. He wrote the first draft of his expedition report while rooming with Mayr in New York, yet he did not apply Mayr's view of evolution in it. Instead, he adopted geneticist Sewall Wright's caveat to natural selection, which holds that in the absence of predators and food competitors, the genetic makeup of small, isolated populations can "drift" in nonadaptive directions. At least for closely related species, Lack reasoned, "pointless" genetic drift could account for the evolution on separate islands of different Galápagos finches—which he placed into just fourteen distinct species. When these originally isolated species later spread to other islands and came into contact, he proposed that "the different species recognized each other primarily by bill differences, and so kept segregated." Lack conceded in this early paper, "Although this is not a completely satisfactory explanation, it is extremely difficult to see any other reason for the origin, and also for the persistence of the bill differences between these closely related species with their similar food requirements." In reaching this conclusion, Lack rejected Lowe's idea that Galápagos finches possess an unusual tendency either to vary genetically or to hybridize freely.[78] The latter finding brought a highly personal prepublication attack from Lowe in the pages of the British ornithology journal *Ibis*, which then refused to publish the brash amateur's paper.[79] Lack turned instead to the CAS as his publisher. Huxley, in contrast, incorporated Lack's discordant views into his 1942 treatise *Evolution: The Modern Synthesis* by invoking the "Sewall Wright effect" to account for the evolution of Galápagos finches. "The release from competition," he asserted, "has permitted what can only be described as an abnormal variability and multiplicity of forms" on the archipelago generally.[80]

Lack joined the war effort following his return to Britain, using his aerial observational skills to spot enemy aircraft on radar and, in idle moments, adapting the new technology to track migrating birds. He also had ample time to reflect on Galápagos finches. "I reached my conclusion, that most subspecific and specific differences in Darwin's finches are adaptive, and that ecological isolation is essential for the persistence of new species, only when reconsidering my observations five years after I was in the Galápagos," Lack later wrote. "In this, of course, I had a distinguished precedent, as Charles Darwin did not perceive the evolutionary significance of the finches until several years after his visit."[81]

In particular, Lack resolved during his wartime isolation that each type of finch beak *must* confer some competitive advantage. Huxley had asserted as much, as a general principle of natural selection, in his 1942 treatise. In this Huxley followed the rule set forth by Soviet biologist G. F. Gause, "If two or more nearly related species live in the field in a stable association, these species certainly possess different ecological niches."[82] In a brilliant new study of Galápagos finches written during a 1943–1944 winter lull in his war duties, Lack announced, "My views have now completely changed, through appreciating the force of Gause's contention that two species with similar ecology cannot live in the same region." Only the fittest (or best adapted) could survive. Since three species of ground finches live together on some Galápagos islands, Lack now asserted, "There must be some factor which prevents these species from effectively competing" against each other.[83]

Although he had not recognized it when he visited the archipelago during the breeding season, when rain falls and food abounds, Lack now saw that the distinguishing beaks of these ground finches must adapt them for eating different seeds (at least during the dry season). Based on scant evidence that he scavenged from the scientific literature, Lack boldly declared, "In the case of the three species of *Geospiza*, there are similarities, but established differences, in their diets, and though further evidence is much needed, it is provisionally concluded here that, so far from being unimportant and purely incidental, these food differences are essential to the survival of the three species in the same habitat." He applied similar analysis to other overlapping finch species.[84] Two decades earlier, after hastily observing that "birds utterly dissimilar in relative proportions of mandibles were feeding upon identical food," Beebe had dismissed the Galápagos as "a mad country for birds." Lack could not accept such madness anywhere in nature. Every living thing must everywhere follow the same rules—Darwin's rules.[85]

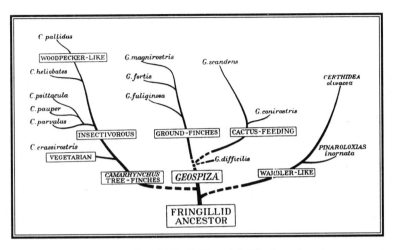

Evolutionary tree, from David Lack, Darwin's Finches *(1947)*
(Source: Reprinted courtesy of Peter Lack.)

A complete picture of finch evolution emerged in Lack's classic book, *Darwin's Finches,* whose title forever fixed a new common name on this family of birds. Galápagos finches no longer, these birds became Darwin's finches in name and understanding. "That Darwin's finches are so highly differentiated suggests that they colonized the Galápagos considerably ahead of the other land birds," Lack wrote. "The absence of other land birds has had a most important influence on the evolution of Darwin's finches, since it has allowed them to evolve in directions which otherwise would have been closed to them." Free from the constraint of competition with mainland birds, some of the island finches remained ground feeders, others took to the trees for food, and yet others adopted the feeding habits of warblers or woodpeckers. From common ancestors, each type evolved on separate islands into different breeding populations in a classic evolutionary process known as *adaptive radiation*.[86]

Thus far, Lack had said little new. Isolation had created these island variations just as it had for Galápagos mockingbirds and tortoises. Unlike those two well-understood groups, however, these finches did not remain isolated on separate islands. They spread into each other's territory. Here Lack broke new ground in his interpretation. "When two related bird species meet in the same region, they tend to compete, and both can persist there only if they are isolated ecologically either by habitat or food," he postulated, ignoring his earlier emphasis on species recognition in maintaining their isolation. Such encounters further push the evolutionary development of both competing types until they diverge into noncompeting types or one dies out. "All of Darwin's finches are isolated from each other ecologically," Lack proclaimed, by diet if not by habitat. Even where they live in the same place, they do not functionally live together. "The evolutionary picture presented by Darwin's finches is unusual in some of its details, but fundamentally it is typical of that which I believe to have taken place in other birds," Lack concluded, "so that with these birds, as Darwin wrote, we are brought somewhat nearer than ususal 'to that great fact—that mystery of mysteries—the first appearance of new life on this earth.'"[87]

Lack's eloquent analysis returned the Galápagos Islands to the center stage of evolutionary science. Yet this resurgence had to await the end of World War II, when another generation of scientists arrived to test Lack's ideas. In the meantime, war-imposed paper shortages postponed publication of Lack's book until 1947, and the spread of hostilities in the Pacific drove scientific expeditions from the archipelago. Things changed after the war for both Lack and the Galápagos. "His work on the Galápagos finches gave David Lack world fame," Mayr notes. "There is no modern textbook of zoology, evolution or ecology which does not include an account of his work."[88] Without having any formal graduate training in science, Lack received a faculty appointment at Oxford University and nearly every honor conferred by biologists. "So at the age of 35," the amateur bird watcher observed, "I became a professional ornithologist."[89] Galápagos science gained renewed status with him.

PART THREE

ECOLOGY MATTERS

SEVEN

BRAVE NEW WORLD

WORLD WAR II UTTERLY TRANSFORMED the Galápagos. As the largest islands in the eastern Pacific, situated only 850 nautical miles from the Panama Canal, they were seen by American military strategists as a potential crossroads of the sea war should the Japanese navy attack the west coast of the Americas. The military carefully prepared for a strike that never came.

Even before the United States joined the war against Japan, naval intelligence began reconnoitering the archipelago. As early as 1936, Victor von Hagen, then in the Galápagos to promote conservation efforts, assisted a secret U.S. Navy inspection of the islands.[1] A series of other navy visits followed, including President Roosevelt's highly publicized one in 1938. By 1941, the American military had charted dozens of bays suitable for harboring ships, submarines or seaplanes and identified Baltra (formerly South Seymour) Island—a small, flat outpost of Indefatigable—as the best site for a permanent airbase.[2]

On December 11, 1941, just four days after the Japanese bombed Pearl Harbor, United States troops from the Canal Zone occupied the Galápagos Islands with the grudging consent of the Ecuadorian government. Naval intelligence prepared a massive field monograph about the archipelago featuring aerial photographs, detailed maps and an analysis of the estimated 810 people then living on the islands, particularly the odd assortment of European nationals residing at Academy Bay.[3] Construction began for an airbase on Baltra in February 1942, with workers dynamiting a mile-long landing strip out of the volcanic rock. They eventually built

over 200 wooden structures for some 1,000 Americans stationed on the island. "The islands that had been a naturalists' paradise became in war a
sunset home for soldiers & sailors," *Time* magazine reported. "For the
G.I., Seymour was The Rock—the never-never land of igneous boulders
and shifting red dust, the U.S. Army's beachhead on the moon."[4] For settlers on nearby Indefatigable and more distant Chatham, the American
military presence meant cash for their fish and produce. Soon the archipelago's human population began to grow.

Building and maintaining a military installation disrupted Baltra's fragile natural environment. The Smithsonian Institution lobbied for a small
laboratory at the base to study the island's endangered species, but netted
only a general order directing military personnel not to shoot the animals.
Apparently no one objected if the bored GIs conducted iguana races or
taunted the feral goats. Recalling his trip to the archipelago, Roosevelt intervened with a memorandum to his secretary of state in 1944, "I have been
at this for six or seven years and I would die happy if the State Department could accomplish something" to protect the Galápagos as an international wildlife sanctuary.[5] This too had to wait until peacetime.

Baltra never recovered from the war. In *Galápagos: World's End*, William
Beebe had described the place as home to a subspecies of particularly large
land iguanas—"veritable dragons," some up to five feet long. They became
prize trophies from his expedition, with the picture of one as the book's
frontispiece.[6] Other prewar naturalists also remarked on the large iguanas
of Baltra without ever suggesting that the absence of smaller, younger
specimens might indicate that the feral goats had already disrupted the island's ecology. As late as 1942, the navy's field monograph noted that land
iguanas "abound" on Baltra. When American forces evacuated the island
in 1946, however, apparently not one of the native iguanas remained alive
on the island—though some in their subspecies survived on North Seymour thanks to members of a Hancock expedition transporting them
there during the 1930s.[7]

Seeking to promote wildlife conservation on the Galápagos, later naturalists made Baltra's iguanas into an object lesson for environmental protection and depicted trigger-happy American GIs as the culprits. Reporting to
the International Union for the Conservation of Nature (IUCN) on a 1954
visit, for example, Irenäus Eibl-Eibesfeldt told of finding only one iguana

specimen on Baltra—a bullet-ridden carcass. "This isle of land iguanas had been wasted for it served during the Second World War as a military base," the German scientist lamented.[8] Never mind that the carcass could not have dated from the American occupation, or that introduced goats and airport construction probably destroyed the iguanas' habitat. The point remained that humans had wiped out Baltra's iguanas. "Is it not a sad commentary on human intelligence," Eibl-Eibesfeldt asked in a magazine article about Galápagos wildlife generally, "that these strange creatures, which still decorate our world like exotic flowers, are within a few years of being completely destroyed?"[9]

Eibl-Eibesfeldt's visit as part of a German scientific expedition to the South Seas reflected a profound postwar development in biology: the rise of ecology. Certainly some naturalists had previously understood organisms in terms of their overall biological and physical environments, but the concept of scientifically studying entire ecosystems came of age only in the mid-twentieth century. With a rising human population, the breakdown of its geographical isolation through air travel and a distinctive, well-documented fauna, the Galápagos Islands offered an ideal place to study ecology, particularly the impact of humans on a near-virgin environment.[10]

Eibl-Eibesfeldt went to the archipelago in 1954 aboard the oceanographic survey ship *Xarifa*. The twenty-five-year-old ethnologist later earned an international reputation for his study of human behavior. On Baltra, he was struck by the ecological impact of people and introduced species on the natural environment. "Nothing," he wrote, "could be more melancholy." What U.S. soldiers had not destroyed, introduced house mice would. "Mice will be last inhabitants of South Seymour, and when they have eaten up all the remaining vegetation, then they will die off. The island will be as bare and barren as it was when it first rose from the ocean's depths and will be a constant and ongoing witness to how, in a few short years, life that had flourished for thousands of millennia could be swept away as a victim of mankind's baleful conflicts." Although biologists criticized Eibl-Eibesfeldt for this simplistic analysis, it reached a wide audience through his popular writings.[11] He expressly absolved the soldiers from moral blame for their acts because publicly blaming people was not this postwar German's way; changing their behavior was. Environmental education became his goal.[12]

Upon his return to Germany in 1955, Eibl-Eibesfeldt sent a memorandum to the IUCN in Brussels urging an international effort to save the Galápagos. "I drew a realistic picture of the situation," he later commented, "and stated that only the establishment of a biological station with a permanent warden could, in the long run, provide effective protection." Station personnel could study the remaining wildlife, enforce existing conservation laws and educate the public about environmental protection.[13] "This seems likely to succeed, judging by the reactions of people with whom we came in contact. They were all proud of their islands even though they did not realize what extraordinary animals they had around them," Eibl-Eibesfeldt reported. "When such things were explained to them, they were willing to learn more and seemed ready to help in the task of preserving these precious species."[14] His appeal to the IUCN reached the sympathetic ears of Julian Huxley.

Huxley had first expressed an interest in Galápagos science during the 1930s, when he championed an international effort to protect island wildlife and sent David Lack there to study finches. As a staunch neo-Darwinist, Huxley readily accepted (and helped inspire) Lack's ultimate conclusion that interspecies competition drove the evolution of these birds. In his 1942 treatise *Evolution: The Modern Synthesis*, for example, Huxley had followed Lack's initial surmise that "non-adaptive" genetic drift in their island isolation accounted for most evolutionary differences among the Galápagos finches. Yet in a 1953 monograph, *Evolution in Action*, he drew on Lack's later work to cite those same finches as a classic case of interspecies competition.[15] This view of the Galápagos as a living laboratory for studying natural selection gave Huxley a professional interest in protecting its wildlife. As the grandson of T. H. Huxley, Darwin's confidant and champion, Julian Huxley already had a historical interest in setting aside the archipelago as a living memorial to Darwinism.[16]

By the 1950s, however, Huxley was less a historically minded scientist than a guru for what he called "evolutionary humanism."[17] He had abandoned academia in 1935 for the first of several highly visible administrative posts. Huxley used these posts as pulpits to preach his gospel of Darwinism and its meaning for humanity. Branching out into new forms of mass communication, he joined the BBC's popular "Brain Trust," a live television program whose panelists fielded obtuse questions from the general public, and supervised the production of nature films for an English movie company, winning an Oscar for his efforts. Having a celebrated grandfather and two famously brilliant brothers (novelist Aldous Huxley and Nobel laureate Andrew Fielding Huxley) enhanced his image. Huxley gained such a reputation for intelligence that one public opinion poll ranked him among the five "best brains" in Britain. His compulsive publishing, lecturing and media appearances, however, always served a larger purpose. "To Julian Huxley," historian Daniel Kevles notes, "it was incumbent upon mankind . . . to struggle to see that evolution continued to flourish. Embracing that aim, human beings could forge an evolutionary ethics, which would start with the principle that it was right to realize ever-new possibilities in evolution."[18]

Unlike his grandfather, who eschewed religion and saw evolution as directionless change, Julian Huxley turned his science into a nontheistic "religion without revelation" that worshiped natural selection as the guiding process. "Searching desperately as a young man for a faith to substitute for Christianity," science philosopher Michael Ruse explains, "Huxley found it in progress—and for him, progress was best manifested in and made most probable and plausible by the evolutionary process."[19] Although this process once operated blindly through brute competition in nature, Huxley believed that humans, by acquiring consciousness through evolution, had gained the power to channel the selection process for their collective benefit. "The general evolutionary battle has been won," he proclaimed in 1964, at the international symposium that formally dedicated the Charles Darwin Research Station on the Galápagos Islands. "As an outcome of Darwin's work, we have begun to grasp the central and all-important idea that man is the latest and highest type produced by the evolutionary process, and that his destiny is to guide its future course on this planet." Scientific research in the archipelago could help show the way to universal evolutionary progress.[20]

This ambition for Galápagos science, which Huxley began promoting as early as the 1930s, grew out of his larger faith in evolutionary biology, and science generally, as lights for human progress. This belief was severely tested by the Second World War. Germany, which still enjoyed a reputation as the world's most scientifically advanced nation even after Nazis had gutted its universities, used quasi-scientific ideas, partly derived from Darwinism, to justify its crimes against humanity; and victory finally came to the United States through a scientific weapon capable of obliterating all people. Throughout, Huxley clung to his belief in science as a progressive force. At war's end, he participated in reviving, as part of the United Nations, a moribund League of Nations agency devoted to global harmony through international education and cultural cooperation. This time, however, Huxley insisted on adding science to the props for global progress—and so he became known as "the man who put the 'S' in UNESCO": the United Nations Educational, Scientific and Cultural Organization. He then served as UNESCO's first director-general and expanded its mission to incorporate nature conservation by founding the IUCN as an affiliate agency, which then spawned the World Wildlife Fund (WWF) to support IUCN efforts to protect endangered species. This international triad became Huxley's principal legacy to Galápagos science.[21]

Huxley articulated a philosophy for UNESCO that inspired its work on the Galápagos Islands. Rejecting traditional religions and ideologies for its guide, he wrote in 1947, "The general philosophy of Unesco should, it seems, be a scientific world humanism, global in extent and evolutionary in background. . . . Such a philosophy is scientific in that it constantly refers back to the facts of existence. It is the extension and reformulation of Paley's Natural Theology and those other philosophies which endeavour to deduce the attributes of the Creator from the properties of his creation." Unlike Paley's theistic moralizing, UNESCO would "relate its ethical values to the discernible direction of evolution, using the fact of biological progress as their foundation." Drawing on his understanding of the evolutionary process as the self-serving actions of individual organisms driving progress for the whole, Huxley concluded that through educational programs, scientific research and cultural exchanges UNESCO should "promote a social organization which will allow individuals the fullest opportunity for development and self-expression consonant with the persistence and the progress of society."[22] The conservation of wildlife and study of ecology fit neatly into Huxley's vision, provided they retained a human orientation.

Conserving native animals in the Galápagos Islands, Huxley maintained, would serve the most profound needs of humans—needs that fundamentally distinguish people from other animals. These needs included "the preservation of all sources of pure wonder and delight, like fine scenery, wild animals in freedom, or unspoiled nature; the attainment of inner peace and harmony; the feeling of active participation in embracing and enduring projects, including the cosmic project of evolution. It is through such things that individuals attain greater fulfilment."[23] The elevation of these quasi-spiritual feelings to the status of basic human needs reflected Huxley's religious commitment to evolution. For him, a Galápagos national park open to eco-tourists and scientists alike could serve both as a source of pure wonder and as a means of understanding "the cosmic project of evolution." In a visionary 1960 report to UNESCO that anticipates the rise of eco-tourism, Huxley stressed that "one of the primary purposes of a National Park is to cater to the enjoyment and interest of visitors." Adventurous visitors from more developed regions would attain spiritual fulfillment, hosts in less developed regions would gain an economic incentive to preserve nature and both sides would gain global understanding through their interaction.[24]

Huxley distinguished between enjoying nature and exploiting it. "The conservation point of view really implies a new ethical attitude as between man and nature," he wrote. "We cannot treat any natural objects merely as things. We must treat them as part of an essentially vital, self-transforming whole."[25] Through UNESCO and other international agencies, Huxley endeavored to foster the study of evolution and ecology in special habitats like the Galápagos, conserve wildlife there and educate people everywhere about them. No one can understand institutional developments in Galápagos science since World War II without appreciating Huxley's view of conservation and its role in promoting evolutionary progress. "By conservation," he wrote in a training memorandum for UNESCO in the year it founded the Galápagos research station, "I do not intend merely preservation, but conservation of resources for present and future use and enjoyment."[26]

Largely independent of efforts by Huxley, Eibl-Eibesfeldt and other Europeans, several Americans also adopted the cause of a science facility on the Galápagos Islands. At the close of World War II, the quasi-governmental Pacific Science Board—then led by Harvard University conservationist Harold Coolidge—recommended using part of the abandoned American airbase on Baltra as a research station. The archipelago's "barren coasts and islands afford simplified ecological conditions, comparable to those of Arctic islands, and provide a veritable field laboratory in themselves," the board noted. "The biological interest of these islands is so great that conservation measures, under the control of such a research station, are urgently required."[27] Ecuador requested the decommissioned base for its own military purposes, however, and so the board's proposal died. Little else happened on the American side until a zoology graduate student from the University of California, Robert I. Bowman, took up the issue following a 1952–1953 research trip to Academy Bay.

Bowman went there to study birds in the wake of David Lack. He had read *Darwin's Finches* in 1949 for a graduate seminar under the botanist G. Ledyard Stebbins. As one of the main architects of the neo-Darwinian synthesis along with Huxley and Mayr, Stebbins was all for Lack's reinterpretation of the data to find interspecies competition for food driving the evolutionary process.[28] Bowman was not so sure. Certainly the beaks of these finches had evolved in response to the available food supply, just as Lack asserted for gross differences between species: small bills for cracking soft seeds, large bills for crushing hard seeds, thin bills for spearing insects, sharp bills for poking fruit and so on. Yet these different kinds of food often existed on the same islands, as did the various species of finches. Lack ultimately attributed subtle distinctions between closely related types of the same or similar species (which typically did segregate by island) to competition for food between types, while maintaining that the proximity and similarity of the islands prevented such fine-tuning from occurring in direct response to the food itself.[29] Bowman disagreed.

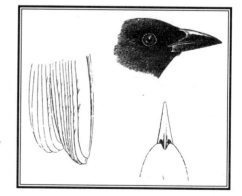

*Cactus finch, from Sclater and
Salvin's report (1870)
(Source: Used by permission of UGA.)*

"Food, food, food," Bowman declared. Differences in food supply alone could explain all the differences among the finches, with each type perfectly adapted to its individual habitat by nature's selection of the fittest individual finches for survival from among its type. No two islands really had the same food supply anyway, he stressed, even if they had similar kinds of food. The distance separating Lack and Bowman may seem small—both invoke natural selection to drive the evolution of Galápagos finches—but Bowman shifts the locus of competition from between types to within them, and thus differently explains their origin. He would later describe Lack's reliance on competition between species as a "British idea" associated with strict Darwinism; he would defend an environmentalist view more in line with such California evolutionists as LeConte, Jordan and Swarth.[30]

The issue became the subject of Bowman's doctoral dissertation. First he reanalyzed the finch specimens at the California Academy of Sciences and elsewhere. When he considered the overall shape of the bill rather than simply a few measurements, he found highly distinct forms where Lack had not. "There is no 'perfect gradation' between genera and between species" as some earlier naturalists had claimed, Bowman asserted, but clear and distinct types. He dismissed the few intermediate forms as chance hybrids. Bowman then went to the Galápagos for five months during 1952–1953 to study the finches in their habitats during both a wet and dry season so that he could see for himself what they ate and whether they competed. He found what he was looking for: "marked botanical and physiographical differences between the various islands," with each form of finch ideally fitted to its niche in nature. "Under the guidance of natural selection the numerous 'adjustments' in bill size and shape have been 'coordinated' with the particular array of available food in such a way as to utilize each adjustment with the greatest mechanical efficiency possible," Bowman claimed. Further, even in the depth of the dry season, he observed different species of finches harmoniously living together without any sign that even the most closely related types vied with each other for food in a manner that could have evolutionary significance. "Since there is no direct evidence that competition is occurring at the present time," Bowman concluded, in defiance of Lack, "I see no logical reason to assume that it must have occurred in the past."[31]

Bowman's work split evolutionary biologists on a critical fault line. "A pitched battle began between the competition camp and the non-competition camp," science writer Jonathan Weiner says of the episode in his book *The Beak of the Finch*, "and in the absence of hard evidence and actual observations it dragged on and on." Both sides called for more research.[32] Based on his recent experience in and around the growing community at Academy Bay, Bowman, like Eibl-Eibesfeldt, worried about the survival of native species on the Galápagos. In 1955, even before finishing his dissertation, Bowman independently wrote to the IUCN pressing for an international effort to save the archipelago's endangered wildlife, and called for a permanent science station to facilitate research there. As an American with recent field experience on the islands, Bowman soon found himself pushed to the front on this issue.[33]

The letters from Eibl-Eibesfeldt and Bowman reached the IUCN at an ideal time to get a response. Huxley was, of course, enthusiastic from the start. Further, Coolidge had recently stepped down as vice president of the IUCN to lead its new Commission on National Parks, created in 1954 to stimulate the establishment of wildlife preserves in places such as the Galápagos. In his worldwide efforts to save parkland, Coolidge worked closely with two other wealthy Ivy League conservationists familiar with the archipelago—Fairfield Osborn, who had taken up where his father had left off in leading the New York Zoological Society, and S. Dillon Ripley of Yale, who had sailed through the Galápagos aboard a 1937 yachting expedition to the South Seas and would later take the helm at the Smithsonian. Then leading the IUCN in Europe, Belgian naturalist Victor Van Straelen saw the archipelago as a new target of opportunity for wildlife preservation just when the end of colonialism threatened his lifetime efforts to establish national parks in his country's African domains. While still in their twenties, Eibl-Eibesfeldt and Bowman joined these venerable conservationists in an international campaign to transform the Galápagos into an Ecuadorian national park monitored by a scientific research station.[34]

The IUCN used the plight of endangered species and the approaching centennial of Darwin's *Origin of Species* to focus public attention on the archipelago. Ripley flew to Quito to brief Ecuadorian officials early in 1956. At its annual meeting later that year, the IUCN adopted a resolution (later endorsed by the Pacific Science Congress) deploring "the precarious situation of various species of fauna and flora endemic to the Galápagos Islands" and urging that UNESCO send qualified naturalists to conduct a survey. UNESCO officials in Europe chose Eibl-Eibesfeldt to lead it. From the American side, Coolidge, Ripley and Osborn arranged for Bowman to accompany him. "They did not want the Europeans to have the whole show," Bowman later explained.[35]

In a coup for the cause, Osborn persuaded the editors of *Life* magazine to send along famed photographer Alfred Eisenstaedt and illustrator Rudolf Freund. Their article, published in *Life* to commemorate 100 years of Darwinism, helped to spawn a new series of highly popular books, the "Life Nature Library," whose first volume opened with a pictorial essay on the archipelago titled "A Showcase of Evolution."[36] At the time, Bowman could still maintain that "few areas of the Western Hemisphere are so little known and so unappreciated" as the Galápagos—but not for long.[37]

The UNESCO expedition surveyed the archipelago for nearly five months in 1957, using an Ecuadorian navy patrol boat to reach almost every island. "The first and foremost fact we were able to verify was that good specimens of every characteristic species of the Galápagos Islands still exist," Eibl-Eibesfeldt noted.[38] Yet profound dangers remained. "Since the airbase opened on Baltra," Bowman warned, "increasing numbers of immigrants have been arriving on these desolate shores. Native wild animals, undisturbed for thousands of years, are threatened with extinction as a result of the clearing of primordial forests, uncontrolled hunting, and the introduction of domesticated animals."[39] In their formal Survey and Report for UNESCO, Eibl-Eibesfeldt and Bowman recommended setting aside nearly all undeveloped land on the archipelago as an inviolate nature sanctuary, strictly protecting all native vertebrates except fish and establishing an international science station to study and guard the wildlife. Further, the survey stressed the importance of educating local settlers through pamphlets and school programs about native species and nature conservation. UNESCO, of course, specialized in such matters.[40]

As the site for the research station, the UNESCO team picked a location on the south side of Indefatigable Island near the existing town at Academy Bay. In a separate article, Bowman described this crude settlement as "about two hundred people living a meager existence plagued by food and water shortages, tedious trails, and uncertain communications with the mainland."[41] The official survey, however, expressed confidence that the town's diverse international population could fully service a scientific research station. "The Seymour air base is nearby," it added, and "at the same time Indefatigable is one of the less spoiled islands rich in natural beauty. Giant tortoises have survived in fairly large numbers, marine iguanas are abundant, flamingoes fish in the lagoons and land iguanas are found in the remote bays of the island." Other biologically interesting islands lay close at hand. In short, the site offered a central perch for watching the entire archipelago at a prudent distance from the main Ecuadorian settlement and administrative capital on Chatham.[42]

The survey placed the archipelago's growing population at 2,000 people—more than twice the prewar total—most living in such poverty that they all but had to plunder the environment to survive. In his private writings, Eibl-Eibesfeldt denounced these settlers for slaughtering tortoises as food and selling penguins as pets.[43] "On the whole colonization conflicts with nature protection. How can both interests be reconciled?" the survey asked. It was a rhetorical question; Julian Huxley had already given the UNESCO answer, which the survey parroted precisely. "Ecuador is interested in developing its tourist trade," the survey observed. "The Galápagos Islands can become an import asset for attracting tourists, but only if they are preserved and protected as national game reserves, like those in Africa."[44]

Life carried the message of a threatened Eden in a glossy cover story calculated to entice a new generation of nature-loving tourists to the archipelago. "Since Darwin's momentous visit," it began, "many of the Galápagos islands have been despoiled. Yet on some, the primeval populations, though waning in numbers, still survive. . . . Nowhere else on earth have reptiles engendered such prodigies as the giant tortoise and the dragonlike marine iguanas. Even the birds have hatched anachronisms." Glorious color illustrations depicted exotic birds, reptiles and sea animals harmoniously coexisting. "Until now man's invasions of the Galápagos have been happily infrequent. It is this circumstance which has preserved aboriginal life and endowed the Enchanted Isles with the look of eternity, as though the river of time had frozen in some peaceful epoch of the prehistoric past." Closing with a call to conserve wildlife, the article stated that the UNESCO mission gives "hope that Darwin's living laboratory of evolution may not perish."[45]

The international effort to save the Galápagos for science and humanity gained momentum. Huxley publicly embraced the survey recommendations by writing in the *Unesco Courier*, "I hope very much that a permanent Biological Station may be set up on the Galápagos with a view to the scientific study and active conservation of their native animals and plants."[46] The appropriate international science organizations fell in line behind UNESCO and the IUCN to support the proposal, including in 1958 the influential International Congress of Zoology, which acted in response to a plea written by Bowman and read to the assembly by Ripley with supporting comments by Coolidge and Huxley. Bowman called on the scientific community to establish "a biological station on the Galápagos Islands in order to promote effective protection of and research on native species" and on Ecuador to "modify laws to permit better protection of Galápagos species." Early in the century, the California Academy of Sciences, Rothschild and others had sought to save Galápagos species for science by collecting them as museum specimens. By midcentury, scientists joined conservationists in repudiating such efforts in favor of saving endangered species for study in the wild.[47]

In 1959, on the one hundredth anniversary of the publication of *On the Origin of Species*, UNESCO created the Charles Darwin Foundation for the Galápagos Islands for the purpose of organizing and maintaining a research station in the archipelago. "No more fitting memorial could be erected to Charles Darwin," Van Straelen proclaimed. Huxley, Van Straelen and Bowman served among the foundation's initial officers; Coolidge and Ripley represented the United States on its executive committee. Despite the dominance of conservationists from Western Europe and the United States, the Ecuadorian government endorsed the foundation's efforts in 1959 and formally set aside the undeveloped regions of the islands as a national park. With early funding from UNESCO, WWF, the American government and the New York Zoological Society, the station's first buildings began rising along the east shore of Academy Bay in 1960. For the first time, the Galápagos Islands had a permanent home for science and conservation.[48]

More than ever, the IUCN-UNESCO campaign focused attention on the archipelago's threatened animal species. These peaceable, photogenic creatures proved irresistible subjects for the new family medium of broadcast television, which in the late 1950s was struggling to refine its programming. Nature documentaries became popular fare, especially on networks experimenting with color programs, and few places could top the Galápagos for easy-to-shoot wild-animal footage. Around the *Origin's* centenary in 1959, for example, television crews from Britain, Canada, France, Sweden and the United States crisscrossed the archipelago. Walt Disney Productions dispatched a team to film footage for its acclaimed nature series, "True-Life Adventures," and several independent filmmakers visited the archipelago, most notably French photographer Christian Zuber, who made two movies and countless still photographs during expeditions in 1958 and 1961. The publicity built so fast that some conservationists worried about tourists arriving in mass before protective measures could take effect. Huxley warned of "the danger of premature publicity, and consequent urgency of getting the Station established."[49]

Dozens of popular books and articles about the islands appeared during this period as well. Never before or since has so much publicity focused on the archipelago. A British ornithologist then working on two remote Galápagos islets concluded, "Film-makers, photographers, more or less professional writers and broadcasters, shoe-string university vacation explorers and a handful of individual adventurers form one heterogenous assembly of odd-corner wanderers, liable to turn up anywhere." He even had a visit from the conservation-minded Prince Philip of Great Britain.[50]

Many conservationists felt that popular accounts of wondrous Galápagos reptiles, birds and sea mammals advanced the cause of nature protection, even among people who could not be lured to visit the archipelago. Huxley and Osborn argued that wildlife films and photos inculcate a sense that people should observe nature, not despoil it. Such images, they believed, could also arouse viewers to support preservation efforts and promote an appreciation of nature. "I am very anxious to get a 'propaganda' film made about the whole question of conservation of wildlife and natural habitats," Huxley wrote in 1960, adding later in a letter to UNESCO that he hoped "by way of propaganda (newspapers, posters, radio, films,

Sketch of waved albatross mating dance, from Frederick
Drowne's diary in Novitates Zoologicae *(1899)*
(Source: Used by permission of UGA.)

etc.) to make people aware of the interest and importance of wildlife in National Parks."[51]

The Galápagos Islands provided a particularly attractive setting for nature stories and documentaries. "Successful nature writers such as Sally Carrighar, Rachel Carson, and Joseph Wood Krutch had softened the violent hues of nature's palette in the postwar years through narratives that drew attention to the interdependent web of life," historian Gregg Mitman writes. For a world that had seen enough killing during the war, wildlife filmmakers were careful to present a more idyllic picture of nature. Mitman notes, for example, "Disney shunned the graphic violence of lions maiming humans and allusions to primal sexuality common in [prewar] travelog-expedition film. Explicit scenes of predators killing prey were rare in the True-Life Adventures."[52] As a land where few predators lurked and most animals showed no fear of people, the Galápagos Islands offered ideal subjects for peace-loving postwar nature writers, filmmakers and photographers.

"In the Galápagos, mockingbirds perched on our hats, and hawks accompanied us on hikes," a Disney filmmaker marveled. An owl once stood "no more than eight feet away, watching us with the intent curiosity of a child." In his narration for a 1959 BBC documentary, WWF leader Peter Scott told of swimming with iguanas, penguins and "the tame turtles in the sea." Swedish nature photographer Sven Gillsäter added, "Our first meeting with the Galápagos Islands' own sea lion also showed us that we were as near to an animal paradise as it is possible to come. A female sea lion was lying suckling her latest infant in a small round hollow in the lava, and she let us come quite close." With the struggles implicit in natural selection carefully edited out, such scenes attracted nature lovers to the archipelago and taught lessons that conservationists wanted to convey. "It was strange to observe this animal curiosity and to find ourselves accepted by the wildlife of this animal land," the Disney filmmaker noted. "And because of it, in the same way that one is charmed by the naivete of a child, we reacted with a friendliness of our own."[53] Here was Julian Huxley's utopia: where evolutionary thinking could enlighten conservation practices for the betterment of all life. By muting natural selection, however, the projected image did not fully convey the prevailing scientific view.

The founding of a research station did not immediately redress the ecological imbalance between the archipelago's growing human population and its threatened native species. Indeed, it initially exacerbated the problem by spurring development on Indefatigable Island—by then increasingly called by its Ecuadorian name of Santa Cruz. Work began on transforming the Baltra airbase into a tourist airport; settlers flocked to Academy Bay; and demand rose for agricultural products from the island's fertile highlands. Zuber feared the worst when he returned in 1961 to shoot a second wildlife film on the island in 1961. "UNESCO has brought us dollars! So a great deal has changed since your last visit," an old acquaintance told him at the dock, but "the tortoises are still being killed as before."[54]

Darwin Foundation officer Jean Dorst conceded as much. "The present population is made up of law-abiding and friendly settlers," he wrote diplomatically in 1961, "but like many elsewhere, they are not always aware of the imperatives of nature conservation. Furthermore, they have thought fit to introduce animals previously unknown in the islands . . . . Everywhere in the islands the reptiles have regressed, especially the tortoises. In former times they were massacred for their fat; now their eggs and young are devoured by the dogs and pigs."[55] Eibl-Eibesfeldt complained bitterly about this development when he inspected progress on the new station. "Tortoise colonies which we visited on Indefatigable Island in 1957 proved to be slaughtered when we visited the same place in 1960."[56] Saving the tortoises in its own backyard became the station's first conservation project.

Despite appearances, the tide had begun to turn for Santa Cruz tortoises by 1961. When Zuber reached the station, he found modest buildings for workers but impressive pens for baby tortoises. "Here, no wild pigs are snapping them up, and these at least would survive," he observed. In an effort

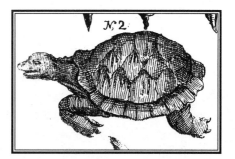

Galápagos Land Tortoise, from Edward Cooke, Voyage to the South Sea *(1712) (Source: Used by permission of UGA.)*

largely implemented by Galápagos-born conservation officer Miguel Castro, station personnel had begun counting the remaining tortoises on the island (eventually finding over a thousand of them) and killing the wild pigs. "We finished these few days of film-making full of enthusiasm for the Charles Darwin base," Zuber concluded, calling it "essential to the preservation from total destruction of the archipelago's plant and animal life."[57] During a separate photo shoot, Gillsäter focused on a different facet of the station's local conservation outreach. "It is UNESCO," he wrote, "which is behind the most important task. Once a week, the school children of Santa Cruz come charging through the cactus thickets for a three-hour lesson on biology, a subject which they never heard of before. . . . The childish enthusiasm has an effect in their own homes, and I think I dare maintain that a giant tortoise has been killed for the last time, at least on Santa Cruz, the island where this animal is still relatively common."[58] To fulfill its mission as a center for science and conservation for the archipelago, however, the station would have to tackle more than small projects on Santa Cruz. Here Bowman, Coolidge and the Americans reentered to give the new station a kick start in 1964.

They called it the Galápagos International Scientific Project, but it was purely an American enterprise. It started out much like any one of a dozen other quasi-scientific expeditions sent to the Galápagos during the early 1960s under the auspices of some zoo, natural history museum or university. In her capacity as an administrator for the liberal arts wing of the University of California extension service in 1962, Alice Kermeen hit on the idea of sponsoring a trip to the Galápagos for students and staff. She naturally contacted the curator of birds at the California Academy of Sciences, Robert Orr, who suggested Bowman as the trip's leader. Having just returned from his third expedition to the islands, Bowman saw an opportunity to promote conservation on the Galápagos at a time when the research station desperately needed a boost. He drew in his well-connected mentor in Galápagos conservation matters, Harold Coolidge, and the two-bit extension trip grew into the largest scientific expedition ever to reach the archipelago.[59]

The size and scope of the Galápagos Project reflected a fundamental institutional development in American science during the postwar era: federal money. Except for agricultural and military research, the United States government had provided little direct financial support for science prior to World War II. That changed during the war, when academic scientists showed their worth by producing atom bombs, radar and countless other military items. Federal spending for scientific research jumped tenfold from 1940 to 1945, to half a billion dollars per year, making the United States government the largest single sponsor of science in the world—a position it never relinquished.[60]

When the fighting ended in an uneasy standoff with its wartime ally, the Soviet Union, the U.S. government continued to support science at ever-increasing levels. In 1950, it established a civilian National Science Foundation (NSF) to distribute federal funds to worthy nonmilitary scientific research and education programs. Maintaining America's economic, political and cultural position in the postwar world, many political leaders believed, required such efforts. "My attitude toward scientists," the powerful Senate appropriations committee leader Richard B. Russell observed at one congressional hearing, "is pretty much the attitude of the boy living in the country going to the country doctor. He thinks the doctor can do anything." Federal appropriations for the NSF rose steadily from one quarter of a million dollars in 1951 to over one third of a billion dollars in 1964, the year of the Galápagos Project. By then, total spending for science by the U.S. government topped $15 billion per year, or roughly 100 times the entire national budget of Ecuador.[61]

As the perennial executive director of the Washington, DC–based Pacific Science Board, Coolidge knew how to tap NSF largesse for research

in the Pacific region. He transformed the University of California's Galápagos trip into an international science and conservation project, enlisting
as its codirector University of California entomologist Robert L. Usinger,
who then served as Pacific Science Board president.

At a critical planning meeting with Kermeen, Orr and Usinger during
the summer of 1962, Coolidge laid out the project. Regarding participants,
the meeting's minutes report, "Dr. Coolidge thought that they should
come from as many countries as possible, including a large number of
Latin Americans." This fit the NSF's cold war mission of fostering international cooperation in science as a way to link the United States with
other non-Communist countries. "Assuming we have fifty participants,"
he suggested that the division "should be about thirty scientists to twenty
advanced graduate students, the 'future doers' who would presumably be
stimulated by this experience into further Galápagos research." This appealed to the NSF's further goal of encouraging science education. Recognizing the NSF mandate to support the full spectrum of scientific
disciplines, Coolidge urged the inclusion of specialists from biology, geology, oceanography and meteorology. "A very few prominent scientists
(Huxley, for example) should nevertheless be invited for political reasons,"
he added, advising that "a list of scientists we plan to invite should be included by name in our application to NSF."[62] Catering to NSF elitism, the
project's application floated such big names as Huxley, Lack and the legendary neo-Darwinian geneticist Theodosius Dobzhansky—but of these
and other noted scientists listed as "expected to participate," only Huxley
joined the project, and his advanced age prevented him from doing anything more than delivering a send-off address for the expedition in California.[63] In the end, over half of the participating scientists came from the
Bay Area.

The Galápagos Project gave Ecuador its first taste of American-style
big science, and it liked the flavor. "Our expedition will be, by at least ten
times, the largest ever to go to the Galápagos," Usinger boasted to the
press.[64] The fifty participating scientists sailed to and from the archipelago
aboard the 425-foot training vessel *Golden Bear*, on loan from the California Merchant Marine Academy, in an arrangement brokered by Coolidge.
For the party's five-week stay in the Galápagos, the NSF built lodging and
laboratories that became vital permanent additions to the research station's

facilities. The U.S. Navy dispatched a support ship with two helicopters to ferry scientists from Academy Bay to various research locations throughout the archipelago, including the first trip ever to the top of tiny, steep Culpepper Island. Journalists and photographers covered the operation from start to finish.

Needing a focal point for the high-profile expedition, Coolidge staged a formal dedication of the five-year-old Charles Darwin Research Station. U.S. Air Force jets flew in 107 dignitaries from the mainland, including two top officers from Ecuador's ruling military junta, six foreign ambassadors and Charles Darwin's great-grandson. The California Wine Institute contributed ten cases of champagne for the occasion, leading one journalist to observe, "This unexpected festive drink, virtually unknown on old Indefatigable, without doubt made this the finest social event in Galápagos history." Reportedly, every local resident turned out for the celebration. Ecuadorian strongman Marcos Gandera Rodriguez was so carried away at the dedication that he then and there pledged his government's full support for the Darwin Station's conservation efforts. To consummate the relationship, project and station leaders flew to Quito, where they negotiated an agreement assigning responsibilities for wildlife conservation to the Darwin Foundation and participated in the issuance of a pair of official decrees protecting native species. For the first time in its history, the Ecuadorian government recognized the archipelago's economic value as a national park. Coolidge, Bowman and Usinger even received medals from junta leaders for their efforts.[65]

The final report submitted to NSF by project organizers proclaimed the expedition a success. "The world-wide publicity resulting from the Galápagos Project and the dedication ceremony has brought the Galápagos Islands and the problems of nature conservation thereon to the fore," the report declared. Further, the report hailed "the tremendous good will generated among the people of Ecuador by the expedition members and the surge of interest by Ecuadorians in their own scientific potential."[66] Like Beebe's expeditions four decades earlier, the trip's main goal was P.R.— this time for the cause of scientifically informed conservation practices. A speaker at the project's initial symposium stated bluntly that, because their "principal attractions are thriving populations of native animals in their natural settings," now "the key to prosperity in the Enchanted Isles is an immediate, effective program for conservation of native wildlife and its habitat."[67] Unspoiled nature for international science and eco-tourism was the project's message.

In their rush to create an image of fervent scientific research in a place where previously there had been little more than collecting, project organizers emphasized style over substance. Past expeditions to the Galápagos had served science mainly by gathering rare specimens—but this offended modern conservationists almost as much as hunting. Under Coolidge's hand, the Galápagos Project foreswore collecting endangered species and large animals; he wanted to demonstrate that scientists could conduct meaningful research on location without the wholesale plunder of native birds, reptiles and sea mammals. Five weeks on the islands, however, was not long enough to do much more than collect material for later study. Participants who were free to gather specimens, such as entomologists and botanists, did the best work—most notably Usinger, who returned with a quarter million insects, and CAS research associate Ira L. Wiggins, whose 1,000 Galápagos plant specimens became the basis of the islands' first truly comprehensive published flora.[68]

Having assembled the expedition in a few months, the organizers drew on a limited pool of scientists. Like Usinger and Wiggins, most of the participants had no previous experience in the Galápagos. Organizers cared more about promoting conservation than doing science anyway, and this called for parading threatened species before the public. Thus famed birdwatcher Roger Tory Peterson went along to see and be seen with the islands' famous flamingos. He did not even pretend to conduct any research. A California medical school professor generated a few news stories by injecting iodine into iguanas, ostensibly to discover something about goiters in humans. Another scientist garnered attention by feeding tiny thermometers to the giant tortoises in a quest to measure stomach temperature during digestion. One reviewer of the expedition's collected papers got the message. Although the papers read like "proposals for research" rather than finished work, the reviewer wrote, they bring "home our ignorance of Galápagos ecology with disturbing clarity" and the urgent need to protect it. This was the project's main legacy to science.[69]

The visible success of the Galápagos Project helped create the scientific, legal and social framework to transform the archipelago into a model site for field research, wildlife conservation and eco-tourism. These developments impressed Eibl-Eibesfeldt during his inspection of the Darwin Station in 1965. "Visiting the Islands now for the fourth time in a period of twelve years, I was able to observe certain changes resulting from colonization," he observed. Earlier, conditions always had steadily worsened for the wildlife, "but now the tide seems changing. Not that the settling activity has stopped, on the contrary: The village of Academy Bay is flourishing." New facilities had sprung up, creating a truly cosmopolitan resort. "I observed the first indication of a change when landing at the main pier of Santa Cruz Island," he wrote. "Although many people were moving around, young and halfgrown marine iguanas basked on the rocks without any sign of fear. This was different from what I saw six years ago. At that time children used to hunt the iguanas for pleasure and they became rare and shy as a result." The German ethologist explained, "A change took place in the people's attitude toward the animals. They were made aware of the iguanas' peculiarity due to an education campaign" by the Darwin Station. Now, from the moment tourists docked at Academy Bay, they would find what they came to see: an animal paradise. Further, Eibl-Eibesfeldt added, "More and more the Station attracts visiting scientists from all over the world." The once-critical investigator hailed these developments as "a most promising start" but warned that too much of even a good thing could destroy the place.[70]

EIGHT

THE HOUSE THAT
LACK BUILT

FROM THE STANDPOINT OF THE HISTORIAN of ideas," the Darwinian anthropologist and immensely popular science writer Loren Eiseley observed in 1968, the Galápagos "are actually the most famous islands in the world. . . . To no other oceanic isles in times known to us have come an equal number of men of the caliber of Dampier, the buccaneer navigator, Charles Darwin, voyager-naturalist, and Herman Melville. . . . Each man saw inevitably with the eye of his time; each had his progenitors, each, in his way, also lent the unique color of his mind to what would pass beyond the observation of a few active volcanic cones projecting from the sea floor. Instead, there was destined to arise a gigantic new mythology of the shaping of the entire creation." This new mythology was the theory of organic evolution, Eiseley explained, and its meaning is "still in the process of elaboration."[1]

These words opened *Galápagos: The Flow of Wildness*, a two-volume set in a prizewinning series of folio-sized picture books about premier wild places published by the conservation-minded Sierra Club. Just as the editors of *Life* magazine inaugurated their pictorial Life Nature Library with a volume featuring the archipelago, the Sierra Club included these volumes in its glossy new series (two of only three set outside the United States). The Galápagos thus gained a prominent place on American coffee tables. Eiseley was the ideal author for the volume: his words could pic-

ture a place as grandly as the finest photographs. He had also written the
era's best-selling history of evolutionary thought, *Darwin's Century*, which
popularized the legendary role of Galápagos finches in inspiring the Darwinian revolution.[2] Now Eiseley repeated the tale of the "astonishing variety of modifications" among those finches, "particularly in their beaks,"
that ultimately led Darwin to his new theory of origins.[3] The *Life* book
said much the same, with pictures of finch beaks illustrating the story of
Darwin's visit to the archipelago. "It was only after analyzing his collection
of Galápagos finches," the text maintained, "that he seriously began to
question special creation."[4]

Actually it was David Lack, during the 1940s, who finally unlocked the
evolutionary significance of these finches and raised them to prominence.
Darwin had invoked Galápagos mockingbirds, not finches, to make his
case in *Origin of Species*.[5] After Lack, however, standard biology textbooks
(which had never before mentioned Galápagos finches) began featuring
extended accounts of their evolution illustrated with charts showing their
adaptive radiation from a common mainland ancestor. Inevitably these
texts followed Lack in calling them "Darwin's finches."[6]

In these accounts, the finches' history reflected the fundamental tenets
of neo-Darwinian evolution. A single ancestral type reaches the Galápagos and spreads out. Individual birds begin modifying in relative isolation
on separate islands, with the fittest forms surviving to create distinct varieties or even species that then spread across the archipelago. Where these
modified forms come in contact, competition for food drives them to diverge still further, accelerating the process of speciation. Lack's view of the
process was textbook dogma by the 1960s. Darwin's finches "shout the
truth of evolution," proclaimed Edward O. Wilson, the author of numerous authoritative books on the topic.[7]

The *Life* and Sierra Club volumes showed much more of the islands
than their drab finches. The *Life* book depicted "the strange subworld of
the Galápagos" where prickly-pear cacti grew into 30-foot trees, tortoises
took on gargantuan form, iguanas learned to swim, cormorants lost the
ability to fly and most animals knew no fear of humans. "Descending from
a few stranded ancestors and cut off from the rest of the world, the Galápagos animals offer much more obvious proofs of the fact of evolution
than can be seen in the more intricate complexes of life in most environ-

ments," the text commented.[8] In his introductory essay for the Sierra Club set, Eiseley went beyond the evolutionary peculiarities of island species to their fundamental unity with life everywhere. "It is the working of such mysterious principles as adaptive radiation and selection which binds this assemblage of extraordinary plants and animals together and relates even the story of man to island tortoises and flightless birds," he affirmed. "We are all, in fact, the product of islands, visible or invisible. At some point in the fossil past, isolation and mutation have diverted each bit of life down some solitary road from which there is no turning back." Oceanic islands simply made this process easier to see.[9]

Eiseley's comments about the significance of the Galápagos for evolutionary research reflected a widely held view. With an international research station fully operational there and generous funding available in Western countries for science, the pace of Galápagos fieldwork began accelerating in the late 1960s and never again slowed down. "No area on Earth of comparable size has inspired more fundamental changes in Man's perspective of himself and his environment than the Galápagos Islands," Robert Bowman noted in 1983.[10] Especially since the Darwin Station's founding, more scientists have studied the Galápagos than any other oceanic archipelago except Hawaii. The emerging pattern of their research reflects the scientific significance of islands under the evolutionary worldview.

As a product of ongoing volcanic activity, the Galápagos Islands present issues of intense interest to geologists. With the acceptance of plate tectonics during the 1960s to explain the evolution of the earth's features, the archipelago became a test case for studying how a single subterranean hot spot produces a chain of islands on moving surface plates. Applied to the Galápagos, this model shows the hot spot generating younger islands through ongoing volcanic activity at the western edge of the archipelago, while the moving plates carry the older islands east. Radiometric dating of rock from various islands confirms this view, and that all the islands are geologically young. The eastern ones also show greater effects of erosion and subsidence, leading to speculation about still older, submerged islands in the chain yet further east, toward the mainland, which may have provided stepping-stones for ancestral organisms to reach the archipelago and more time for native species to evolve.[11]

Island view, from James Colnett's report on Pacific whaling (1798)
(Source: Used by permission of UGA.)

The emergence of new islands from the sea presents living laboratories
for field biologists, especially if those islands are so situated as to receive
only a few terrestrial species from the mainland and are not fouled by humans introducing foreign organisms and exterminating native ones. An
isolated group of several islands adds to the evolutionary dynamic. The
Galápagos exhibits these ideal features better than any other place on
earth. With the Darwin Station to aid visiting scientists and the inaugu-
ration in 1969 of scheduled air service to Baltra, biological fieldwork on the
Galápagos exploded.

Biologically speaking, oceanic islands share certain hallmark characteristics. First, they typically have fewer kinds and a narrower range of plants
and animals than continents—plenty of birds, perhaps, but scarcely any
land mammals. They also often harbor some native species that live
nowhere else, making them "endemic" to the islands. Biologists thus speak
of a "disharmony" or a discrepancy of species represented between an island and the nearest mainland. Further, species that can cross seawater
(such as winged animals and plants with windblown seeds) predominate
on islands and commonly appear in grossly modified forms from kindred
mainland types. These island forms often lack defensive adaptations and
generally seem tame, which renders them highly vulnerable to invaders.
Especially in archipelagos, a particular type of organism may take on a variety of island-specific forms, each neatly adapted to local conditions. The
question ultimately arose, why these kinds of forms in these kinds of
places? God could have created such distinctive, fragile creatures for islands, of course, but that struck Darwin as irrational. He found his answer
in organic evolution, and accepting that premise as their starting point, his
present-day disciples return to islands with further queries.[12]

"The two basic questions in evolutionary biology are (1) how does evolution occur and (2) why does evolution occur?" notes Princeton University biologist Peter R. Grant, the most respected of current Galápagos researchers. "By studying evolution in the relative simplicity of island environments we hope to be able to answer both questions using the full range of techniques of modern biology: description, experimentation, modeling, statistical analysis, and the ordering of data to make inferences about evolution in the past."[13] Since 1973, Grant has led annual research expeditions to the Galápagos searching for answers to these questions. In any given year during that period, ten to twenty other groups of scientists from universities across the globe make the same pilgrimage. Most of them study some signature characteristic of island biology as applied to Galápagos species, seeking some incremental insight into the grand pattern of evolution. They typically fly into Baltra's modern airport and scatter by boat to various field sites throughout the archipelago, relying on the Darwin Station to provide logistical support. Their cumulative work has generated thousands of scientific articles about the evolution of Galápagos species.[14]

"Ever since Darwin it has been believed that studies of speciation on islands may throw light on the general processes of species origin," the great geneticist Theodosius Dobzhansky explained. "Of course, there is the possibility that island speciation may be a special case, different from speciation of widespread continental species. Be that as it may, studies of speciation on islands are a fascinating chapter of evolutionary biology." So too speciation on the Galápagos may be a special case, different even from how it operates on other islands and archipelagos, but it has nevertheless fascinated scientists and inspired evolutionary studies generally.[15]

Botanist Duncan Porter has studied endemism and disharmony among Galápagos plants, building on the work of Ira Wiggins and Yale Dawson. He has catalogued some 550 types of vascular plants native to the islands, over two-fifths of which exist nowhere else on earth, in addition to another 200 types introduced through human contact. "The indigenous flora is full of taxa particularly well adapted to crossing the sea barrier between the archipelago and the mainland," he finds, whether by wind, on water or in birds. "It is a disharmonic assemblage of species adapted to long-distance dispersal." Porter sees no need to invoke a local creation to account for the peculiar Galápagos flora. Nearly all of the native plants came from South America, he concludes, and were carried by chance to the Galápagos, where many of them evolved into something new. A handful have gone on to generate multiple forms.[16] Of these few, one of the most famous is the prickly-pear (or *Opuntia*) cactus, which has fourteen distinct island forms, including the world's largest type with a woody trunk supposedly evolved to protect it from grazing tortoises. It is giants versus giants in the Galápagos, the biologists postulate. Observing the massive rafts of vegetative debris floating down the broad Guayas River of western Ecuador, Dawson once wrote, "It is not difficult to visualize a piece of cactus—fruit, seeds or stems—moving slowly westward with the Humboldt Current toward the Galápagos Islands." This, he believed, could be the source of the many Galápagos cacti, though Porter credits their introduction to transoceanic birds.[17]

Those floating mats could also carry rodents, or so zoologists assert. Seven species of small rice rats are the only land mammals native to the Galápagos except migratory bats, which apparently could fly there. Darwin collected the first specimens of these rats in 1835, and evolutionary zoologists have studied them ever since as a classic case of long-distance mammalian dispersal followed by adaptive radiation. Indeed, Deborah Clark notes, "The Galápagos native rats hold the world record for sea crossing by terrestrial mammals."[18] Analysis of their anatomy, tissue and blood suggests that these seven species, each found only on the archipelago, evolved over time from a few chance immigrants. All but two have become extinct in the years since European and American ships carried black rats to the place—another textbook example of island species succumbing to continental competitors. For most types of mammals, zoologists find smaller, weaker forms native to islands and larger, stronger forms native to continents; predicting the winner is easy in a struggle for survival between them.[19]

"Whereas mammals tend toward dwarfism on islands, reptiles tend toward giantism," notes science writer David Quammen. "The list begins with those huge tortoises. It includes both groups of endemic Galápagos iguanids, the marine iguana *Amblyrhynchus cristatus* and the land iguanas of the genus *Conoloplus*."[20] Except perhaps for Darwin's finches, these giant tortoises and iguanas are the Galápagos' best-known creatures. Their sheer size, adaptation to island life and survival from an earlier reptilian age still fascinate evolutionary scientists just as they once intrigued Darwin. Chief among present-day researchers, Craig MacFarland and Thomas Fritts specialize in the archipelago's tortoises, while Howard Snell devotes his career to iguanas. They have learned more about Galápagos reptile behavior than anyone before them, and each ultimately became formally associated with the Darwin Station, which works especially hard to save these particular species from extinction.[21]

"The living was superb," MacFarland says about his first two-year stint amid Galápagos Tortoises during the late 1960s. "We found that the giant tortoise leads a generally peaceful, lazy life," mostly sleeping, grazing and basking in the sun. MacFarland discovered a symbiotic relationship in which the small ground finch cleans the tortoise of ticks, and confirmed various aspects of breeding behavior useful to the conservation effort. A graduate student at the time, MacFarland trekked with his wife and newborn child through some of the islands' most remote regions, studying nine of the eleven surviving tortoise populations. Each population constitutes a distinct race, isolated on a separate island or highland and readily recognizable by the shape of their shells. Those from certain low, arid islands have peculiar saddle-back carapaces. All others have the standard dome. Saddlebacks can reach their heads up for scant vegetation in dry places but forfeit the ability to fully retract into their shells—presumably a fair trade-off in a land historically devoid of predators, and yet another lesson in natural selection.[22]

Iguanas are the islands' other famous relict from the reptilian era. Just as only on the Galápagos have some tortoises acquired saddle-back shells, so too only there have some iguanas taken to the sea. Taxonomists divide Galápagos iguanas into three separate branches, with two closely related terrestrial species in one genus and one distant marine relative in another—though keen observation by Howard Snell and his wife, Heidi, has detected some breeding across these branches.[23]

The unique marine form attracts the most attention from visiting scientists. Irenäus Eibl-Eibesfeldt first visited the Galápagos to observe the behavior of marine species in 1954 and kept returning to learn more about it. Dee Boersma took up this work during the 1970s in a five-month study

Galápagos Marine Iguana, from Charles Darwin, Journal of Researches, *2d ed. (1845)*
(Source: Used by permission of UGA.)

that documented marked differences in adult body sizes among the various island populations of marine species—a finding fraught with evolutionary significance. Further research by Martin Wikelski and Fritz Trillmich linked these differences to natural and sexual selection. Hailing marine iguanas as "diving champions of the reptile kingdom," scuba pioneer and naturalist Jacques-Yves Cousteau sailed his world-famous research ship *Calypso* to the islands during this same period to record their actions on film and study their adaption to deep-sea pressure.[24]

Inevitably, researchers investigating Galápagos reptiles are struck by their seeming tameness. MacFarland changed his baby's diapers on the backs of sleeping tortoises. Trying to shoot an action film, Cousteau complained that "perhaps the most salient characteristic of the life-style of marine iguana is its inactivity." Researchers must step around them on shore and can easily swim with them in water. Cousteau's divers fed them by hand underwater. "It is generally believed that the marine iguana never accepts food from man," he reported, "but here, in the open water, . . . we succeeded." After painstaking efforts to reintroduce land iguanas to Baltra Island from neighboring North Seymour during the 1990s, Snell found that they would not budge for airport buses; now his concern is that Baltra drivers learn to share the road with iguanas. Of all the earth's creatures, only humans care about saving the endangered ones. For their part, Galápagos reptiles stubbornly retain their signature indifference to people. In a land historically without predators, no endemic animal instinctively fears becoming prey.[25]

Such striking indifference spurred Darwin's thinking on the effects of isolation and evolution on wild animals. He rode on land tortoises but could not guide their direction; he repeatedly tossed marine iguanas into the sea only to watch them return to their perches on shore. They did not seem to care about him at all. They were not tame but simply indifferent—a quality not limited to the islands' reptiles. "A gun is here almost superfluous," Darwin observed, "for with the muzzle of one I pushed a hawk off the branch of a tree." This defenseless behavior allowed Darwin to build his collection of Galápagos land birds, which provided fodder for his thoughts about evolution through isolation. "There is not one which will not approach sufficiently near to be killed with a switch, and sometimes, as I have myself tried, with a cap or hat." They would learn caution through contact with settlers, Darwin predicted, but this form of evolution on the Galápagos has not proceeded very far to date.[26]

The fearlessness of Galápagos birds continues to astound visiting scientists and facilitate their research. Working on two remote Galápagos islands during the mid-1960s, for example, the Scottish ornithologist Bryan Nelson was able to conduct an unparalleled study of seabird behavior. With his wife, he camped among thousands of breeding boobies, gannets and frigate birds. "For a year we were Adam and Eve; there was nobody else on earth, so far as we could tell," Nelson fondly recalled, though in fact a sizable contingent from the Galápagos International Scientific Project visited them, and Prince Philip dropped by to film a conservation documentary for British television. Nelson could never have observed seabirds on the mainland so closely without capturing or killing them. He walked among them in the Galápagos without ruffling a feather, and learned about their breeding and feeding behavior, which he found to "testify to the all-powerful influence of natural selection, in the neo-Darwinian sense, in producing the wonderfully adapted forms that live lives, often far from enchanting, in the enchanted islands."[27] Yet just as in Darwin's day,

*Sketch of young boobies, from
Frederick Drowne's diary in*
Novitates Zoologicae *(1899)*
(Source: Used by permission of UGA.)

it is Galápagos land birds that provide the most telling evidence for evolution from the archipelago. In particular, science writer Jonathan Weiner
asserts, Darwin's finches provide evolution "we can watch."[28]

"The Grants are leaders in this field," Weiner says of the finch research
conducted by Peter Grant and his wife, Rosemary. They migrate annually
to tiny Daphne Major to study the same finch population over time,
recording that which changes and that which doesn't. "This is one of the
most intensive and valuable animal studies ever conducted in the wild,"
Weiner notes. "It is the best and most detailed demonstration to date of
the power of Darwin's process."[29] Daphne Major is so small that the
Ecuadorians didn't even bother to give it a new Spanish name when they
rechristened the archipelago's islands; it is big enough, however, to serve as
home for resident breeding populations of several hundred medium
ground finches (*Geospiza fortis*) and a lesser number of cactus ground
finches (*Geospiza scandens*). Largely undisturbed by mockingbirds or humans, Daphne Major is the ideal roost to watch these finches evolve.[30]

At first, the Grants neither concentrated their attention on Daphne
Major nor anticipated making its finches their career. The stimulus to
restudy Darwin's finches—by then so well-known from the work of Lack
and Bowman—came from postdoctoral student Ian Abbott in 1972, when
Peter Grant was a thirty-six-year-old biology professor at Canada's
McGill University. By then the British-born Grant already had an international reputation for doing precise field studies on the role of competition in the evolution of birds and rodents. His earlier research, particularly

with rock nuthatches in south-central Eurasia, questioned the actual impact of character displacement in the evolutionary process—that is, whether the difference between two similar species increases when they live together due to competition between them.[31]

This issue lay near the heart of the dispute between Lack and Bowman. Neither doubted that the gross differences among Darwin's finches (such as between ground finches and tree finches) resulted from their differing hereditary adaptations to the diverse array of island environments open to them in the absence of most other land birds. Yet Bowman saw this process as sufficient for their total development right down to the finest adjustments of their beaks, whereas Lack invoked competition among species to account for the narrower gradations between them. Lack's view had reigned for two decades, but by the early 1970s it was under assault from a new generation of biologists such as Bowman, who stressed the physical environment's role in shaping species.[32] Ian Abbott developed a scheme to test the matter among Darwin's finches on the Galápagos, and Grant (who was already interested in the birds because of their extreme variability) signed on as his mentor.[33]

"Darwin's finches are the most suitable group of island birds for addressing these issues," the Grants observed. These finches are highly variable, easy to handle live and found in varying combinations largely undisturbed by humans.[34] Further, ever since Lack, leading evolutionary biologists had singled them out as textbook examples of character displacement: for instance, in the definitive articulation of the concept, W. L. Brown and E. O. Wilson in 1956 cited an example of Galápagos medium and small ground finches taken directly from Lack. "On most of the islands, where they occur together, the two species can be separated easily by a simple measurement of beak depth," Brown and Wilson asserted. Yet on tiny Daphne Major and Crossman Islands, where the one or the other lives alone, "a sample of ground finches gives a single unimodal curve exactly intermediate between those [two types] from the larger islands."[35] Ernst Mayr drew on similar evidence from Lack as his lead example of the concept in the 1963 treatise, *Animal Species and Evolution*, though he calls it *character divergence*—a term he borrowed from Darwin, who described the process as "of high importance" to natural selection.[36] Having already disposed of one "classical case" of character displacement in his analysis of nuthatches, Grant welcomed the chance to reexamine Darwin's finches; it became the study of a lifetime.[37]

In that first year, the Abbotts and the Grants studied all manner of Galápagos ground finches at eight widely scattered sites on seven different islands, including Daphne Major. They explained their choice: "The six species in the genus *Geospiza* were selected because they are widely distributed in the Archipelago, several species occur together on some islands, they are abundant in the accessible coastal regions, and they show most strikingly the very subtle gradation in beak sizes both within and among islands."[38] Peter Grant worked in the field with Ian and Lynette Abbott. Rosemary handled logistics and cared for the Grants' two young daughters. At each site, the researchers inventoried the seeds and fruit, weighed and measured a random sample of netted finches and observed which species ate what food. After they had been collecting such data for three rainy months early in 1973, Ecuadorian biologist Tjitte de Vries convinced Peter Grant that dry seasons are different for finches on the Galápagos. So the team returned later in the year for two more months, to do it all over again.[39] The compiled data found some middle ground between Lack and Bowman.

Grant commenced a 1977 lecture on his team's findings with a quote from British geneticist J. B. S. Haldane, "There are still a number of people who do not believe in the theory of evolution. Scientists believe in it, not because it is an attractive theory, but because it enables them to make predictions which come true."[40] His team, Grant reported, had found a predictable evolutionary pattern among Galápagos ground finches. Like Bowman, the Grants and Abbotts documented sufficient differences in food supply on the various islands and in feeding behavior among the finches to account for gross divergences in beak sizes and shapes. Big-beaked ground finches ate larger, harder seeds and lived only where such food grew in abundance; small-beaked ground finches ate only the widely available smaller, softer seeds. "All *Geospiza* populations exhibited selection for and against various species of seeds and fruit," the Grants and Abbotts asserted in a joint journal article. "Our data, and those of Bowman, indicate an adaptive significance of the beak differences"—which was more than Lack ever conceded to the environmentalist camp.[41]

Like Lack, however, the Grants and Abbotts found signs of interspecies competition at work in molding finch beaks. "First, the occurrence of particular combinations of *Geospiza* species on islands is non-random," Grant asserted. Except on Daphne Major, only species with markedly different beaks live together—those that do not compete for food. "Missing combinations comprise species which are size-neighbors on the bill-size spectrum," he stressed. "Whenever these species have met, one has been competitively excluded." Further, after close analysis of the data collected in 1973, Grant detected that, on average, the beaks of any paired species of ground finches tended to differ more when they lived together than when they lived apart.[42] Thus Grant finally saw in Darwin's finches what he had failed to see in rock nuthatches—evidence of competitive exclusion and character displacement.[43]

"I have established a case for the role of interspecific competition" in the evolution of ground finches, Grant stated in his 1977 lecture, "but I think the case . . . will be strengthened if it can be shown that directional selection occurs *now*" in response to current changes in food supply.[44] This would require continual monitoring of the finches, not simply a onetime snapshot. Having once lured Grant back to the Galápagos with the promise of seasonal fluctuations in rainfall and food supply, de Vries now assured him that "no two years are alike."[45] El Niño floods and La Niña droughts punctuate long-term weather patterns in the region. Grant closed the lecture by again quoting Haldane: "No scientific theory is worth anything unless it enables us to predict something which is actually going on. Until that is done, theories are a mere game with words, and not such a good game as poetry."[46] With a view of evolution as good science, not bad poetry, the Grants began returning to the Galápagos.

For their ongoing study they chose Daphne Major, that undisturbed islet with the two most similar species of ground finches. Here they hoped to witness the continuing impact of interspecific competition. "Daphne Major is a suitable island because it is small," Peter Grant explained early in the study. "It has manageable populations of about 1500 *G. fortis* and 400 *G. scandens*, which are size-neighbors and hence potential competitors." Working with a succession of students during every breeding season since 1975, the Grants measured, weighed and banded these birds and their descendants. "We have the opportunity of making crude estimates of the heritabilities of [beak and body size] by relating offspring measurements to parental measurements," Grant noted. "We also have the opportunity of following the fates of cohorts to see who survives best and who reproduces most." The test, he predicted, would come when the annual rains failed and food became sparse. After three years of waiting, drought ultimately came with a vengeance.[47]

Peter T. Boag served as Grant's assistant on Daphne that year. He watched the plants wither, food disappear and the finches die. "*Geospiza fortis* did not breed at all in 1977 and suffered an 85 percent decline in population," Boag and Grant reported in *Science*. The slightly larger *G. scandens* suffered a 60 percent loss. Most of the birds simply starved to death, but all did not starve equally. "Small seeds declined in abundance faster than large ones," Boag and Grant noted, "resulting in a sharp increase in the average size and hardness of available seeds." Among the *G. fortis* population, larger birds with deeper beaks could crack the harder seeds and so survived in greater proportions than the others. As males are typically larger than females, they suffered less loss. By 1978, the average beak and body sizes for the population had measurably increased. When the few surviving females then selected the largest, deepest-beaked males for mating, these averages increased further—and these traits were passed along to the next generation.[48] Natural selection resulting from competition for food and sexual selection tied to mating preferences had caused a predictable hereditary change in the ground finches of Daphne Major. "Evolution in action," Jonathan Weiner called it. "No one had ever seen Darwin's process work that fast."[49]

The pendulum swung back during the historic El Niño of 1983, when the Galápagos Islands received record rains. This time Lisle Gibbs stood watch for Peter Grant on Daphne and saw the floods wash away a half decade of directional selection. "Birds bred continuously for eight months, instead of the usual one to three months," Gibbs and Grant reported in *Nature*, "and by the time the rains ceased both food level (seeds) and bird density were exceptionally high." All finches did not share equally in the bounty, however. Small, soft seeds now predominated on the island, transforming it into a paradise for birds with smaller beaks and bodies. "Despite an initially large standing crop of seeds, finch mortality was high and cor-

BILLS OF THE GENUS GEOSPIZA.

Finch beaks, from Walter Rothschild, Novitates Zoologicae *(1899)*
(Source: Used by permission of UGA.)

related with size, possibly because large birds had difficulty finding enough large seeds," Gibbs and Grant noted. The population quickly outstripped the island's carrying capacity, and larger birds suffered most. Larger males died in disproportionate numbers. The average beak and body sizes for the population moved back toward the old norms. "The chief implication of the results of this long-term study is that natural populations subject to the effects of rare climatic perturbations are not at a fixed equilibrium state,"

Boag and Grant concluded. This population adapted to changed conditions at the level of its genes.[50]

"Evolution occurs when the effects of selection on a heritable trait in one generation are transmitted to the next generation," the Grants explained.[51] This happened among medium ground finches on Daphne Major after both the drought of 1977 and El Niño in 1983. If either climatic change had remained constant over time, then presumably it could have pushed the island's *G. fortis* population permanently in a new direction. Indeed, in its major public pronouncement on the factual basis for evolution, America's prestigious National Academy of Sciences highlighted Peter Grant's estimate that "if droughts occur about once every 10 years on the islands, a new species of finch might arise in only about 200 years."[52] As it happened, however, the locally cataclysmic events of drought and El Niño largely canceled each other out and left the local population oscillating between two optimal forms without breaking out into a new type. Nevertheless, in a *New York Times* feature article, Wiener hailed the offsetting episodes as a "handy-dandy evolution prover."[53] Creationists, in contrast, dismissed the finch beak oscillation as meaningless in the long run. "Nothing new had appeared," antievolution leader Phillip E. Johnson of the University of California asserted in a *Wall Street Journal* Op-Ed piece, "and there was no directional change of any kind."[54] More than a century and a half after Darwin first made them famous in his *Journal of Researches*, the Galápagos Islands and its finches continue to serve as a center court shuttlecock in the creation-evolution debates.

After working on the Galápagos with their students each year for over a quarter century, the Grants have done far more than track the oscillations of beak and body sizes on Daphne Major. As their daughters grew older, Rosemary Grant started participating in the fieldwork and research.[55] Together with their students, the Grants have conducted dozens of studies on finches across the archipelago. "The work as a whole," observed Oxford University biologist William Hamilton, "gives the most detailed unified support to the Neo-Darwinian view of evolution that the theory has yet received."[56] Members of the Grant team have intensively monitored the mating of Galápagos finches across species, for example, to gauge the role of hybridization in the evolutionary process.[57] The team recently began using DNA samples to chart genetic relationships among the fourteen branches of the bird's family tree—leading to their conclusion that the warbler finch most closely resembles the ancestral type that originally immigrated to the archipelago.[58]

Most doggedly, however, the Grants have pursued the Darwinian issue of competition among species. This was their primary interest, of course, but the scientific assaults on their initial findings of interspecific competition stiffened their resolve to look for more evidence. Florida State University biology professors Daniel Simberloff and Edward F. Connor led the attack beginning in the late 1970s. They subjected the Grant team's 1973 data on competitive exclusion to sophisticated statistical analysis, only to find that random colonization could largely account for the reported distribution of finch species among the various islands. Even if chance alone did not predict so many missing combinations of similar species, Simberloff and Connor asserted, "competitive exclusion is one, but only one, of several possible explanations."[59] Other statisticians came to the Grants' defense.[60] The notion of competition among species so divides various ecologists and biologists that neither side will readily give ground. Surveying the battlefield early in the 1980s, two researchers wearily concluded that no study could ever finally settle this war, adding "that further efforts in this direction will only sow more confusion."[61] By this time, however, the Grants and their students were back in the Galápagos collecting more ammunition.

Dolph Schluter is the student most closely associated with the Grants' study of interspecific competition on the Galápagos. He began in 1979 by examining the finches of three similar islands on the archipelago's northern periphery for evidence of competitive exclusion and character displacement—two telltale signs of past competition stressed by Lack. Schluter found such evidence most spectacularly on Genovesa, where the sharp-billed ground finch seems to exclude the small ground finch and to take on some of its characteristics, and on Pinta, where these two species neatly subdivided the food supply apparently through character displacement. He accumulated similar evidence from other islands in succeeding years, all of it fitting neatly into a Lackian framework.[62] "I find Lack's intuition really stunning given how little information he had," Schluter says. "He's my hero actually. . . . They should be called Lack's finches."[63] Peter Grant compares Lack's work to a house built on Darwin's blueprint. "It is still standing," Grant observes. "It suffered storm damage in the early 1960s [from Bowman] and late 1970s [from Simberloff and Connor]. This has been repaired." With original construction by Lack and renovations by Grant and company, the modern master of Darwin's finches concludes, "structurally it is the same building, but it stands on a firmer foundation now."[64]

Robert Bowman remains skeptical. He questions some of the data underlying the Grants' arguments and maintains that the physical environment could cause the current diversity among finches. His ongoing research into the role of learned song patterns in mate selection for Galápagos finches further supports his stress on environmental factors in the origin and persistence of specific types. Interspecific competition still strikes him as a distinctly British idea: Darwin, Lack and Peter Grant all graduated from Cambridge. Bowman expresses unshakable confidence that fieldwork in the archipelago will vindicate his position and, as a example of things that he hopes will come, points to a recent proposal to reclassify Genovesa's sharp-beaks as small ground finches. That would kick one prop out from under the house that Lack built. "The Galápagos have long held great interest and fascination for naturalists of all persuasions," Bowman once wrote and still believes. "Today, the truly challenging objections [to neo-Darwinism] come . . . from some of the leading authorities on the theory of natural selection [who] do not doubt that evolution took place or continues to unfold, but rather . . . question the traditional ideas of its *modus operandi*."[65]

Disagreements between environmental determinists such as Bowman and advocates of competition such as Grant constitute a continuing family squabble among Darwinists. Both sides look for evidence from nature to advance their position and see the Galápagos Islands as an ideal source of new insights. Oceanic islands present simple biological communities that "take us back to a point in time when continents were simple," Dolph Schluter explains. Archipelagos are especially useful because they allow field biologists "to test ideas by making comparisons among islands." No other island birds in the entire world are better than Darwin's finches for studying speciation because, as a group, they are young and relatively undisturbed. "They represent an early stage in the diversification of a group, and hence allow us to identify the causes of the origin of an adaptive radiation," Peter Grant notes. "By direct study it will never be possible to achieve the same understanding of the marsupial or dinosaur radiation as of the Darwin's Finch radiation, but the principles established for the finches can be applied to the marsupials and dinosaurs."[66] On these points, Bowman would wholeheartedly concur.[67]

Due to the work of Grant, Bowman and Lack, evolutionists of all stripes now see Darwin's finches as the product of generation-by-generation selection-mediated change in gene frequencies within populations—which leaves this historic avian subfamily largely on the sidelines in the current high-profile debates over whether the neo-Darwinian synthesis can account for all evolutionary processes, including the origin of major groups. Grant readily concedes, "The manner in which fourteen species of Darwin's Finches came into being—an adaptive differentiation initiated in [isolation]—is not universal." Other species may form without isolation or by nonadaptive pathways. Macroevolution, if it exists as a distinct process, may build on contingencies not found in microevolution. In this regard, Grant notes that "there are limits to generalizing from the single example of Darwin's Finches."[68] Partisans in the heated debates over the sufficiency of the neo-Darwinian synthesis rarely invoke Darwin's finches except to show that natural selection does, at some level, produce species. On this there remains widespread agreement among evolutionists.[69]

Creationists, of course, see things differently, even on the Galápagos. Contributors to the *Creation Research Society Quarterly* periodically address the challenge posed by the various forms of Darwin's finches, sometimes by splitting them into many quite closely related species and other times by lumping them into one or two highly variable types. One such contributor spoke for them all when he stated, "The birds are all still finches, and there is no evidence of change of the magnitude which macroevolution would require." Time has not transformed them into "different kinds of birds, such as ducks, hummingbirds, and vultures," another contributor stressed. The Genesis account in the Bible simply declares that God created each basic "kind" (not each narrow species), biblically minded creationists inevitably explain, and so the sort of microevolution displayed on the Galápagos does not exclude reading that account literally. Further, some resolute antievolutionists point out that the Grants have not actually observed even the microevolution of new species, but rather projected that result from inconclusive oscillations in gene frequencies.[70]

Galápagos schoolchildren learn the tenets of creation science at Loma Linda Academy, just down the road from the Charles Darwin Research Station. Loma Linda is one of the finest schools on the archipelago, build by the fundamentalist Seventh-day Adventist Church to serve students of all faiths on Santa Cruz Island. Its pupils include several children of Ecuadorian parents employed by the research station. Since 1970, tourist dollars have attracted tens of thousands of Ecuadorians to work in the Galápagos. These settlers have drawn missionaries to the place, some of whom preach creationism. The Roman Catholic Church, which operates the oldest and most powerful radio station on the archipelago, occasionally broadcasts sermons against Darwinian materialism. Settlers and scientists thus carry the creation-evolution debate to the Galápagos, albeit in muted tones. Dominating Loma Linda's interior courtyard, a building-sized mural depicts Galápagos birds, reptiles and sea lions proclaiming the message in Spanish, "All creation exalts the Creator." Most visiting scientists do not speak that language, and even if they did, would read the statement quite differently from how the Adventists wrote it.[71]

"What kind of God can one infer from the sort of phenomena epitomized by the species on Darwin's Galápagos Islands?" philosopher of science David Hull asks. "The evolutionary process is rife with happenstance, contingency, incredible waste, death, pain and horror." Hull knows his Darwin: for him it is matter only, all the way down. He flags the philosophical problems faced by those who would believe that God created life, individual species and humans through a purely Darwinian process. "Whatever the God implied by evolutionary theory and the data of natural history may be like, He is not the Protestant God of waste not, want not. He is also not a loving God who cares about his productions. He is not even the awful God portrayed in the book of Job. The God of the Galápagos is careless, wasteful, indifferent, almost diabolical. He is certainly not the sort of God to whom anyone would be inclined to pray."[72]

New Yorker correspondent David Denby muses on such matters in an article chronicling his recent tourist cruise through the archipelago. He applied the enforced idleness of the voyage to reading from the shelf of new books on Darwinian philosophy and evolutionary psychology. "It is a remorseless process," Denby observes about Darwinism early in his voyage. "Does anyone—even the most confident atheist or materialist—really take comfort from evolution by natural selection?" By the end of the trip, however, Denby has come to terms with the process. "Like it or not, I was part of the ceaseless flux, which includes crabs, spiders, sea lions, frigate birds, and eco-tourists, all equals in nature," he affirms. "How small our adventure is—and how much even that smallness is to be prized!"[73]

Denby describes the Galápagos of the 1990s as "a kind of eco-tourist Lourdes, visited by more than fifty thousand people a year." These modern-day pilgrims, he surmises, receive spiritual healing in a Darwinian world by swimming with sea lions or watching up close the flight of the magnificent frigate bird.[74] Expressing kindred feelings more than a quarter century earlier, ornithologist Roger Tory Peterson titled his *National Geographic Magazine* article on the Galápagos "Eerie Cradle of New Species," and wrote of "the new breed of traveler who comes to see, to marvel, and not to destroy." This cradle is eerie because, in the Darwinian creation account (as the Grants found among the small ground finches on Daphne Major), evolution comes through the death of the many and the survival of the few.[75]

In his article, Peterson repeatedly compared the Galápagos to Eden, but a more apt biblical comparison is to Ezekiel's Valley of Dry Bones, where the wind breathes new life into dead bodies. Before Darwin, many Galápagos visitors could not so readily imagine life arising naturally from death. Perhaps that is why some, such as Melville, depicted the Galápagos in such ghostly terms, stressing death rather than rebirth. Yet where Melville cast the place as a Dantean purgatory inhabited by damned souls entombed in "self-condemned" tortoises, Darwin hailed it as "a perennial source of new things."[76] Thus in *The Beak of the Finch*, Jonathan Weiner dwells on the significance of death for natural selection. "There is a special providence in the fall of a sparrow," he writes to conclude his depiction of the drought of 1977. "Even drought bears fruit. Even death is a seed."[77]

Those who reject a Darwinian viewpoint still can see a special providence in the Galápagos simply by overlooking the waste and indifference Hull sees there. One glance at the mural in the Loma Linda schoolyard proves this. The Adventists' painter depicts happy creatures harmoniously exalting their creator without a shade of competition or shadow of death. Countless Galápagos settlers and visitors undoubtedly see the island animals in this light. Public opinion surveys find that fewer than 10 percent of Americans accept a strictly Darwinian view of origins; the percentage must be even lower in Ecuador.[78] This surely affects how tourists view Galápagos animals, much as it affected Elizabeth Agassiz when she toured the archipelago with her husband Louis on the *Hassler* expedition more than a century ago. The Agassizes near-pantheistic spirituality helped them to see evidence of God in nature. Her published letters from the Galápagos to the *Atlantic Monthly* tell of sea lions bewailing a death in their number and of diverse species living in idyllic harmony. "The 'happy family,' so often represented in menageries, was to be seen here in nature," she declared.[79] Such prominent evolutionary naturalists as Walter Rothschild, David Starr Jordan and Henry Fairfield Osborn also admired Galápagos species without seeing Darwinian gods of competition bedeviling their interaction. It is modern neo-Darwinians who most starkly confront David Hull's piercing question: Could anyone believe in God and the Galápagos?

David Lack converted from agnosticism to Christianity in the same year that his classic book on Darwin's finches appeared. "Before my book, almost all subspecific differences in animals were regarded as nonadaptive," Lack noted. By dwelling on the selective power of competition and death, he determined that most such differences among Galápagos finches were adaptive—and from him the idea spread, changing how evolutionists view nature.[80] Lack's unshakable commitment to neo-Darwinism cemented his long friendship with Ernst Mayr: "We were both convinced that all phenomena in the living world, particularly adaptations, had to be explained as the outcome of natural selection," Mayr said in eulogizing Lack. Yet about religion, the atheistic Mayr added, "we entirely disagreed." Lack's devout faith baffled many of his scientific colleagues. "He, in a manner I could never understand, was able to embrace within his mind both Darwinian theory and orthodox Christianity, apparently by keeping them in water-tight compartments," fellow Oxford University zoologist Alister C. Hardy observed.[81] Hardy, for his part, labored to find a biological basis for religious experiences and won the coveted Templeton Prize for Progress in Religion for his efforts, but that was not Lack's way.

"Various writers, some Christian and others agnostic, have been troubled about natural selection," Lack once wrote, "because it is so unpleasant. Natural selection works because in each kind of animal most individuals die before they have produced any offspring, while most of the rest die before they have borne as many offspring as they might." More than any ornithologist before him, Lack documented the staggeringly high death rate of birds and used it to explain their evolution. Such findings obliterate Paley's natural theology, Lack conceded, but need not reveal anything about God. "Although on theological grounds the ordering of the animal creation may to some persons seem surprising, man is surely unqualified to judge whether this ordering is in any way evil, or contrary to divine plan," he offered. Science deals with nature, religion with the human soul, and each should remain in its separate sphere. Just as "science has not accounted for morality, truth, beauty, individual responsibility or self-awareness," Lack maintained, "it is also not necessary and undesirable to postulate that animal evolution has been helped by supernatural interferences with natural law." God and science, as Darwinian paleontologist Stephen Jay Gould would later describe them, were "non-overlapping magisteria" for Lack. "Christians have not infrequently become atheists and atheists Christians," Lack concluded. "Either view involves unexplained gaps and contradictions."[82]

Late in life Darwin said of God, "I feel most deeply that the whole subject is too profound for the human intellect. A dog might as well speculate on the mind of Newton. Let each man hope & believe what he can."[83] Lack exemplifies one response to religion among legendary Galápagos researchers—Darwin, Agassiz, Jordan, Beck, Osborn and Huxley suggest others. Drawing a sharp distinction between belief in God and being religious, Ernst Mayr observes, "Virtually all the scientists known to me personally have religion in the best sense of this word."[84] Perhaps Mayr has in mind such a scientist as fellow National Academy of Sciences member John C. Avise, who studies the genes of the giant sea turtles that swim in Galápagos waters. Rejecting any belief in God except the "genetic gods" of neo-Darwinian evolution, and dismissing natural selection as having "no intelligence, foresight, ultimate purpose, or morality," Avise nevertheless affirms, "Human existence is not meaningless just because life has arisen from natural processes and ends quickly for the individual." Meaning can come through evolution, he asserts, and through applying the knowledge that humans have gained about the evolutionary process.[85] Due to their historical and ongoing significance for evolutionary thought and the impact of such thinking on religious belief, the Galápagos Islands stand at the crossroads of science and religion. For scientists and visitors alike, they can be a spiritual place.

NINE

PARADISE IN
PURGATORY

YOU COME FOR THE ANIMALS," American essayist Annie Dillard wrote in *Harper's Magazine* in 1975. "You walk among clattering four-foot marine iguanas heaped on the shore lava, and on each other like slag. You swim with penguins; you watch flightless cormorants dance beside you." She generalized: "The animals are tame. They have not been persecuted, and show no fear of man. You pass among them as though you were wind." A generation that had embraced pacifism and environmentalism during the late 1960s wondered whether such a peaceable animal kingdom really existed. "Walk into the water," Dillard confirmed. "A five-foot sea lion peers into your face, then urges her muzzle gently against your underwater mask and searches your eyes without blinking. Next she rolls upside down and slides along the length of your floating body, rolls again, and casts a long glance back at your eyes. You are, I believe, supposed to follow."[1]

Dillard, then a contributing editor for *Harper's*, first went to the Galápagos in 1974 on a standard tourist cruise. She had just published *Pilgrim at Tinker Creek*, which launched her career as one of America's foremost nature writers. That book, a *Walden*-like journal about wildlife in rural Virginia, contrasted scenes of sublime grace and beauty with horrid waste and death. The former reaffirmed Dillard's conviction that a Creator exists; the latter led her to describe creation as "the brainchild of a deranged manic-depressive with limitless capital." Unlike David Lack, Dillard never kept her science and religion in watertight compartments, and the specter of evolution darkens her post-Darwinian natural theology. "The faster death goes, the faster evolution goes," she observed. "Look: Cock Robin may die the most gruesome of slow deaths, and nature is no less pleased."[2] Lack observed the same forces at work among Darwin's finches while holding God blameless. Dillard could not do so along Tinker Creek, but her view brightened considerably on the islands. She titled her article "Innocence in the Galápagos."[3]

Although Dillard wrote with greater profundity than the typical travel writer, her attention was drawn to the same two features of the archipelago that attract most visiting journalists and eco-tourists. Of course they come for the animals, Dillard recognized: the tameness and strangeness of the native wildlife dominates nearly every visitor's recollections. "The songbirds are tame. On Hood Island I sat beside a nesting waved albatross while a mockingbird scratched in my hair"; even "the wild hawk is tame," she added. "If you take pains, you can walk up and pat it." The rush of sea lions stood out as her most vivid memory. "Theirs is the greeting the first creatures must have given Adam—a hero's welcome, a universal and undeserved huzzah. Go," she commanded, "and be greeted by sea lions."[4] Her readers responded in countless numbers, lured by the prospect of experiencing nature without binoculars or a gun.

Dillard served up the spectacle of evolution in overdrive as the archipelago's other feature attraction, albeit with a twist. "It all began on the Galápagos, with those finches," she reminded readers, presenting as true the legend of Darwin's epiphany. "The finches evolved in isolation. So did everything else. With the finches, you can see how it happened," Dillard explained without ever mentioning the role of competition in that process. "Geography is life's limiting factor. Speciation—life itself—is ultimately a matter of warm and cool currents, rich and bare soil, deserts and forests, fresh and salt waters, deltas and jungles and plains." Perhaps this rendition provided all that most visitors and readers wanted to know about evolution: leave it with benign forces of adaptive radiation working through isolation in differing environments, even if Dillard knew otherwise from her days along Tinker Creek. In the Galápagos, further, she openly sided with the "many," as she counted them, who question the "sheer plausibility" of the modern neo-Darwinian synthesis that reigns in Western science, and longingly suggested that "appending a very modified neo-Lamarckism to Darwinism would solve many problems." For Dillard, "the triplet splendors of random mutation, natural selection, and Mendelian inheritance are neither energies nor gods; the words merely describe a gibbering tumult of materials." She preferred to see a world where "freedom twines its way around necessity, inventing new strings of occasions."[5]

Her treatment of evolution permitted Dillard to remain uncharacteristically upbeat throughout her account of the Galápagos. Literary critics have compared her other work to that of Melville without noting that both authors wrote popular journal articles about the archipelago—one distinctly dark, the other quite light.[6] "I will sing you the Galápagos Islands," Dillard concluded her essay. "It's all happening there, in real light, the cool currents upwelling, the finches falling on the wind." She ended with a quote from Samuel Taylor Coleridge to the effect that, in visiting the Galápagos, she had "drunk the milk of Paradise."[7] Melville's nineteenth-century hell had become the late twentieth-century eco-tourist's heaven.

The Galápagos had been slow to reclaim the attention of foreign visitors following the disruption of leisure travel during the Second World War. Of the prewar regulars, only the American yachter Irving Johnson continued to return after 1945. Almost single-handedly, he kept the place alive in popular travel literature through a steady stream of articles in places such as *National Geographic* and *Yachting*—but few tourists were persuaded to follow him there. "The Galápagos are not inviting," he wrote with his wife in a 1949 article. "Today's small population is composed mostly of Ecuadorian convicts, who don't want to be there, and a tiny band of settlers, who like to be left alone."[8]

Each return visit brought back reports of more settlers living worse lives. By 1959, Johnson described Puerto Baquerizo Moreno, the archipelago's burgeoning administrative capital on San Cristóbal (formerly Chatham), as "a village of shacks" grown threefold in three years. El Progreso, an ill-named nearby farm town, was a "frontier settlement [where] pigs root in the muddy streets and an aura of poverty clings to the dilapidated dwellings."[9] Johnson still enjoyed the island wildlife, but hardly in a manner that raised ecological consciousness. The lack of effective control, he wrote in a 1956 travel guide, "has left the yachtsman an almost virgin territory where he can cruise and explore, hunt and fish to his heart's con-

CHATHAM ISLE, *one of the Galapagos taken at 1 1/2 miles dis!*

Island view, from James Colnett's report on Pacific whaling (1798)
(Source: Used by permission of UGA.)

tent. . . . You can catch all the birds you want by hand. . . . The penguins make good pets."[10]

A solitary article about the archipelago in the tourism benchmark *Travel* during the 1950s confirmed Johnson's bleak assessment. Based on his reading of prewar literature, its author, Frank Rohr, had dreamed of sailing to the Galápagos—but in the wake of the postwar influx of economically distressed Ecuadorians, he found its romance gone. From the first, Rohr saw only desperate settlers and slaughtered tortoises. "If any of us had been entertaining secret hopes as to the beauty and availability of the island girls, this first sight convinced us there were so pitiably little of the former, the latter wasn't worth investigating," Rohr warned readers. "As passionately as we had wished to raise the islands on our voyage out, we were glad to be sailing home again."[11]

Among the growing number of settlers struggling to survive on the Galápagos during the two decades following World War II, only a handful relied on tourism for their livelihoods. Most of these lived near Academy Bay on Santa Cruz Island (formerly Indefatigable) in the emerging village of Puerto Ayora. The Angermeyer brothers, who had fled to the islands from Germany before the war, stood out as the archipelago's chief tourist guides. They knew where to find the colonies of birds, reptiles and sea mammals that visitors most wanted to see—and would take them there for a price. In 1960, an expatriate American sailor named Forrest Nelson opened Puerto Ayora's first restaurant and hotel. With the founding of the Charles Darwin Research Station there, the town soon outstripped Puerto Baquerizo Moreno as the archipelago's chief tourist port.[12]

Hoping to capitalize on the worldwide attention generated in 1964 by the Galápagos International Scientific Project, an Ecuadorian travel agency began offering tourist berths aboard the *Cristóbal Carrier* on its semiregular ten-day cargo runs from Guayaquil to the archipelago's five small settlements and back. "The *Carrier*, a converted wartime L.S.T., is such a battered, rusty old tub that her seaworthiness is something one would not wish to see tested," famed nature photographer Eliot Porter wrote from personal experience. "The enduring redolence of past cargos of dried fish, spoiled produce, and frightened cattle, whose dung was allowed to accumulate on the forward deck and in the hold, permeated every crack and corner even to her superstructure. In the cramped first-class cabins of the upper deck, the stained, grubby bedding, covered with a single sheet, musty from the sweating bodies of dozens of previous passengers, clearly suggested what conditions prevailed in the less luxurious accommodations." Regarding the attractions of such a cruise, promotional literature simply stated, "The Galápagos are among the last major archipelagos of the world that are completely undeveloped," and promised "an experience you will never forget." This offer was accepted mostly by young backpackers, some of whom jumped ship on the unpatrolled islands to camp illegally in the wild. As late as 1969, an American travel magazine could still complain, "All tourists that come to the Galápagos Islands must come by private yacht or on this decrepit cattle boat."[13]

Although it did not recommend visiting the place quite yet, the 1967 edition of the *Encyclopedia of World Travel* informed readers, "the Ecuadorian government has plans to make the Galápagos a health and vacation resort."[14] These plans built on UNESCO's vision of stabilizing ecological and economic conditions on the archipelago by developing a national park to attract environmentally sensitive tourists. "In a world of expanding urbanization and contracting wilderness, natural preserves are especially attractive," one speaker informed the Galápagos International Scientific Project. "A Galápagos National Park would be a mecca not only for naturalists, scientists, students, and philosophers of all nations, but for everyone capable of sensing evolutionary forces in action."[15] Although Ecuador had proclaimed such a park in 1959, government officials did nothing to define, protect or develop it for six years. In 1965, they finally called in a team of British park planners to study the archipelago and recommend "the best means of exploiting its economic potential through tourism."[16]

The planners proposed flying tourists directly to Baltra air base, where they would board luxury tour boats with private cabins, catered meals and a fixed itinerary of the best nature sites. Passengers would remain on board throughout most of their visit, with trained naturalists to guide them at predetermined landfalls. "Experience elsewhere has shown that no matter how much he enjoys the sights of nature the average international tourist tires easily and after a few hours wants to be able to relax in comfort," the planners stressed. "He also wants to enjoy good food and drink and a comfortable bed at night. Shortcomings in regard to any of these makes him disgruntled."[17] These recommendations served as the foundation for future park development.

By 1970, Ecuador had fixed park boundaries to include 97 percent of the archipelago's land, commissioned wardens to patrol them and begun scheduled flights to Baltra on the government airline, TAME. The country's largest travel agency, Metropolitan Touring, pitched in with a sixty-passenger cruise ship that met arriving and departing flights at Baltra in the course of plying a weeklong circuit of island nature sites. Lindblad Travel Agency of New York offered longer cruises at higher prices. In 1970, the first full year of scheduled air service, 4,500 persons visited the Galápagos—a number that marked the start of seemingly endless growth in island tourism.[18]

From the outset, the international consulting firm hired to devise Metropolitan Touring's plan of operations recognized the archipelago's potential. "To the man who knows nothing of science, the Galápagos Islands will be fascinating—a place which is quite unlike anything he has ever seen before," the initial consulting report stated. "To the scientist, or the person with scientific leanings, the Galápagos will be close to paradise." The key to commercial success lay in proper packaging and marketing. "The Islands cannot offer miles of sandy beaches, towns with Spanish flavor, Inca ruins or the life of big hotels," the report noted. "The main reason that tourists will visit the Islands will always be to see the animals." Let them view these animals from a comfortable boat, the report recommended, with stops among the playful South Plaza sea lions, the concentration of penguins and iguanas on Fernandina's fresh lava flows, the seabird rookeries of Española and North Seymour, Isabela's surviving tortoises and the like. Except for the Darwin Station exhibits in Puerto Ayora, the cruises should steer clear of other humans and let the passengers commune with nature among themselves.[19]

Conservationists Gifford Pinchot (left) and Cornelia Bryce Pinchot (right) examine specimens in the Galápagos during their 1929 expedition to the South Pacific. Courtesy of Special Collections, California Academy of Sciences, Pinchot Collection

Templeton Crocker (right) and aide collecting bird specimens during Crocker's 1932 expedition to the Galápagos. Courtesy of Special Collections, California Academy of Sciences

Allan Hancock with land iguana during his first expedition to the Galápagos, 1927. Courtesy of Special Collections, California Academy of Sciences

Zoologist Ernst Mayr, 1954. Photo by
Nathan W. Cohen. Courtesy of Betty
Cohen. Special Collections, California
Academy of Sciences

Julian Huxley, honorary president of the
Charles Darwin Foundation for the Galápagos
Islands, at the time of its founding in 1959.
Courtesy of Robert I. Bowman

Herpetologist Joseph Slevin from the 1905–1906 Academy
expedition back in the Galápagos with Hancock's Oaxaca
(in background), 1927. Courtesy of Special Collections, California
Academy of Sciences

Ornithologist David Lack, 1966. Courtesy of Lack family archive

Cactus Ground Finch, from Charles Darwin, ed., Zoology of the *Beagle, 1838–43. Courtesy of University of Georgia Libraries*

Large Ground Finch, from Charles Darwin, ed., Zoology of the *Beagle, 1838–43. Courtesy of University of Georgia Libraries*

Irenäus Eibl-Eibesfeldt and Robert Bowman at their proposed site for the Charles Darwin Research Station during the UNESCO Galápagos survey, 1957. Courtesy of Robert I. Bowman

Ornithologist Robert Bowman examines Medium Ground finch on the Galápagos, 1986. Courtesy of Robert I. Bowman

Darwin Station researchers Heidi Snell (left) and Howard Snell (right) studying the iguanas of Santa Fê Island, 1980. Courtesy of Heidi Snell

Ornithologist Rosemary Grant adjusts mist net on Daphne Major Island, 1999. Courtesy of Peter R. Grant, photographer

Ornithologist Peter Grant adjusts mist net on Daphne Major Island, 1999. Courtesy of B. Rosemary Grant, photographer

Barren Daphne Major Island, where the Grants have studied finches regularly since 1973. Photo by author

Conservationists Victor Van Straelen
of Belgium (left), Andre Gille of France
(middle) and Harold Coolidge of the
United States (right) at the Darwin
Station dedication, 1964. Courtesy of
Robert I. Bowman

With the Golden Bear behind, Galápagos International Scientific Project
codirector Robert Usinger (right) and Ecuador junta coleader Col. Guillermo
Freile Posso examine marine iguanas during the Darwin Station dedication,
1964. Courtesy of F. Schuke, photographer. Special Collections, California Academy
of Sciences

Members of the Galápagos International Scientific Project on Santa Cruz
Island, 1964. Courtesy of Robert I. Bowman

Ecuador postage stamp of Galápagos Land Iguana, 1999. Courtesy of Heidi Snell

Ecuador postage stamp featuring the tortoise Lonesome George, international symbol of Galápagos conservation, 1999. Courtesy of Heidi Snell, photographer

Ecuador postage stamp of Galápagos blue-footed booby, 1999. Courtesy of Heidi Snell

Crew setting up IMAX 3-D camera atop volcanic Bartolomé Island, 1999. Photo by author

San Cristóbal monument erected in 1935 to mark the 100th anniversary of Darwin's visit to the Galápagos, as seen in 2000. Photo by author

Tourist directions in Puerto Ayora, Galápagos, 2000. Photo by author

Galápagos students under schoolyard mural depicting local animals praising the creator, 2000. Photo by author

"A carefully devised advertising and promotion program will be necessary if the ordinary tourist is to visit the Galápagos," the consulting report advised. "In contrast, the educated person, the professional, or the scientist . . . can be attracted simply by letting him know that a good tour of the islands and adequate facilities are available."[20] To accomplish these ends, tour organizers invited journalists such as Annie Dillard to join island tours—generating a stream of promotional articles from travel writers on the lookout for something new. In the United States alone, fantastic accounts of Galápagos cruises soon appeared in *Esquire, Harper's Bazaar, Vogue, Sunset* and numerous major newspapers. *Audubon* devoted nearly an entire issue to one reporter's ecstatic Galápagos travel diary. The slick travel monthly *Holiday* ranked the archipelago as a "Dream Place of the 70's."[21] Almost overnight, the Galápagos became a chic destination. UNESCO made it official in 1978 by proclaiming the Galápagos National Park as a World Heritage Site, one of the first four natural areas so designated. In a major coup for conservationists and tour organizers, Ecuador established a Galápagos Marine Resources Reserve that limited commercial fishing in interisland waters to licensed local fishermen using traditional methods.

The national park's popularity boosted the economy, of course, but its impact reached deeper. If these early articles reflected visitor reactions generally, then the park promised to fulfill Julian Huxley's original vision of a nature sanctuary that would heighten ecological and evolutionary awareness worldwide. "The mildly adventurous tourist can now walk in the company of Darwin," a Sunday *New York Times* travel article announced in 1970. "This is beyond any question the most exciting place in the world for the tourist who carries with him a curious interest in the world, its origins and its future."[22] The following year, *Harper's Bazaar* hailed Galápagos travel as "one of the great mind-expanding experiences."[23]

Also in 1971, journalist Richard Atcheson (a self-identified jaded New Yorker) described his own consciousness-raising cruise among the islands. "It was the most disarming and moving experience in my life to encounter creatures who were not in the least afraid of me," he wrote in *Holiday* magazine. "I am not St. Francis, and in advancing age have almost lost the dream I once had of emulating him—but in those days on those islands, I felt very Franciscan indeed, overwhelmed with an old fantasy about the integration of all living things, a lying-down-together in wonder." His greatest distress on the trip came when "a fellow-passenger from Arkansas, who had been in all the places I had been and seen (supposedly) all I had seen, assured me over dinner that this evolution stuff was a lot of hokum." How dare a heretic blaspheme on holy ground?[24]

Even as he exalted the experience of Galápagos travel, Atcheson worried about its consequences. Increased tourism would bring more visitors and settlers to a place where defenseless endemic animals had survived solely because of their isolation. Man "has already destroyed countless species" elsewhere, Atcheson said; "now I'm afraid he'll do a job on the Galápagos."[25]

Numbers bore out Atcheson's concern. During the 1970s, the annual count of tourists visiting the archipelago tripled to over 17,500, while the resident population nearly doubled to almost 6,000. These figures seemed certain to grow as the national park raised its visitor limit from 12,000 to 25,000 tourists per year. The number of tour boats mushroomed, threatening visitors' sense of isolation within the park; small hotels, restaurants and souvenir shops sprang up along Academy Bay; and day trips from Puerto Ayora to close-in park sites became a popular option for independent travelers. "In 1968 when I made my first trip to the Galápagos, there was only one car in Puerto Ayora," Smithsonian Institution vulcanologist Tom Simkin complained in 1981. "Now . . . you have to look both ways to cross Main Street."[26] On islands elsewhere in the world, more people mean fewer native species. "Now if you have money enough (and it's a bargain, considering), you too can visit the Galápagos," Atcheson noted. "For the sake of the blue-footed booby, I wish you wouldn't, but how can I in all conscience urge you not to? It is a fantastic trip."[27]

Scientists shared Atcheson's concern. The risks posed by tourism to the fragile island ecosystem were given top priority by the Charles Darwin Research Station in its ten-year plan for the seventies.[28] In 1973, its director joined a small committee of foreign and domestic experts in drafting a Master Plan for the Protection and Use of the Galápagos National Park. The plan set aside certain zones solely for supervised scientific research, some with marked tourist trails for limited mixed uses and others for intensive visitor use. Further, the plan required that tourists visit the national park solely in small groups accompanied by licensed guides, and that their activities be tightly controlled. Park wardens, paid from visitor access fees, would patrol the entire preserve.[29]

Suddenly the archipelago became a laboratory for studying the impact of tourism on a near-pristine environment. The Darwin Station designed park rules, trained visitor guides, analyzed the effectiveness of those rules and guides and educated tourists about appropriate conservation practices.[30] Heavy visitor traffic did damage a few sites, but park officials addressed each discrete problem with added restrictions. On balance, most observers soon concluded, increased tourism actually benefited the archipelago's ecosystem by giving more people an economic or emotional stake in protecting it.[31] Ecuadorian business and government leaders did not want to kill the goose that laid the golden egg of tourism, and Atcheson surely spoke for countless island visitors when he fretted, "Perhaps I should never have gone there, because now I'm worried about the future of the blue-footed booby."[32] Assessing ecological developments on the Galápagos, even Irenäus Eibl-Eibesfeldt finally conceded in 1974, "They are positive and they have surpassed my expectations." Compared to when he first inspected the islands for UNESCO in the 1950s, he noted, "Tourism, although enormously increased, is now under control."[33]

Native animals basked in worldwide attention, especially the big, beautiful or bizarre ones. Of course, saving the giant tortoises came first. Surveys conducted during the 1960s found ten of the original fourteen or fifteen island races surviving in the wild, but only the populations on Isabela (formerly Albemarle) and Santa Cruz appeared capable of sustaining themselves without human intervention. Each race faced a different challenge. A vigorous breeding population lived on Pinzón Island, for example, but alien black rats had killed every hatchling for four decades prior to the survey. On Española, in contrast, feral goats so ravaged the vegetation that scarcely any tortoises survived. Beginning with a small experiment in 1965 and expanding over time, Darwin Station staff collected tortoise eggs from Pinzón for hatching and the few survivors on Española for breeding. Hatchlings raised from Pinzón eggs went back when they grew large enough to fend off rats, while Española tortoises remained at the station until their island was cleared of goats. Similar efforts ultimately reached six island races, saving each from threatened extinction. A writer for *Science* calculated the cost of repatriating Pinzón tortoises at two thou-

Dome-shaped Galápagos
Tortoise, from Albert
Günther's treatise (1875)
(Source: Used by permission
of UGA.)

sand dollars each. "Now, $2,000 for a tortoise might seem excessive in a country with a per capita annual income of $1,200, but it's a bargain by many standards," he wrote. "Last year Americans donated $200,000 so that rubber flippers could be surgically attached to a sea turtle in Florida that had been attacked by a shark."[34]

The Darwin Station found that rescuing tortoises was enormously popular in the United States and Western Europe. Tourists flocked to see the station's brood. The San Diego Zoo contributed a purpose-built rearing center and one of its own Española males. When park wardens discovered an emaciated, middle-aged saddleback on Pinta Island in 1971, it became news around the world. No one had seen a tortoise on the island since the California Academy of Sciences explorers killed three there in 1906. "Probably very rare," they had reported at the time, and rarer still in 1971.[35] Park officials dubbed him Lonesome George and took him for safekeeping to the Darwin Station, where eugenic-minded naturalists, who strictly segregate tortoises by race, put him in a pen of his own. Despite the offer of a $10,000 reward, no mate was ever found. "He made an ideal poster boy for conservationism," *Science* magazine reported. "Lonesome George, the male without a mate, the last survivor of a doomed race." Scientists convened a symposium to discuss his fate and ultimately decided to keep him at the station, where he might live for another century. "If George can be used in a fund-raising campaign to loosen millions of dollars from wallets, it would be very important in saving other species," Galápagos tortoise expert Peter C. H. Pritchard explained.[36]

Saddleback Galápagos Tortoise, from Albert Günther's treatise (1875) (Source: Used by permission of UGA.)

When a five-month wildfire threatened tortoise habitats on Isabela Island in 1985, the world's media followed it closely. Teams with special firefighting aircraft flew from as far away as Canada to help. The park wardens and station personnel who carried dozens of giant tortoises to safety became international heros.[37] As their numbers slowly crept up due to these and other extreme conservation measures, tortoises once again became a common site for tourists and the living symbol of the islands named for them. By the 1980s, a Darwin Foundation officer could boast, "All of the races of giant tortoises known to exist when the [station] was founded (and even two believed extinct at that time) are now safe for posterity, apart from the Abingdon (Pinta) sub-species, in which case the sole surviving male will be the last of his kind unless a mate can be found in some zoo." The Española tortoises remain one of the few animal types ever successfully reintroduced into the wild exclusively from captive breeding stock.[38]

Although the Darwin Station gained fame primarily for its tortoise work, its comprehensive conservation program aided a wide variety of native species and became a model for applying science to wildlife and habitat preservation. For example, as United States Peace Corps volunteers assigned by the Smithsonian Institution to the Darwin Station during the late 1970s, Heidi and Howard Snell perfected a captive breeding program for threatened populations of land iguanas. Defending this use of Peace Corps workers, Smithsonian Institution secretary and longtime Darwin Station board member S. Dillon Ripley offered a purely pragmatic justification: "Quite aside from any scientific information revealed, the Galápagos people will benefit from tourism and the creation of new jobs." Station and park officials often used such economic arguments to defend their conservation efforts, particularly the more costly or intrusive ones, but surely scientists had a more esoteric reason to care so much about Galápagos species: they mostly wanted to watch them live and evolve in an undisturbed environment.[39]

For this goal, the eradication of goats, dogs, cats and other introduced mammals took precedence over settlers' objections. Through their destruction of the natural habitat, feral goats posed the greatest threat to native species and thus became the prime target of conservationists. "The only good goat on the Galápagos," native flora expert Duncan Porter explains, "is a dead goat." By the early 1980s, park service hunters had eliminated them from five smaller islands (killing some 100,000 in the process), and had set their sights on exterminating them from larger ones as well. Park and station personnel also claimed some success in the battle to clear dogs and cats from sensitive bird and reptile breeding sites throughout the archipelago, but less so in their efforts to eradicate alien plants, rats and fire ants.[40]

With goats and tourists coming under control by 1980, many conservationists eyed settlers as the next problem to be addressed. Given all the high-tech planning that went into developing controlled eco-tourism in the Galápagos, some of the initial plans should have anticipated and addressed the potential environmental impact of the settlers that would follow the tourists. None did. Yet the surge in tourism during the 1970s had attracted a flood of job-seeking settlers from Ecuador. Both trends continued through the next decade, with the annual number of tourists doubling to more than 40,000 by 1990, and the resident population reaching 10,000 or more. Ecuador had granted provincial status to the Galápagos in 1973, which prevented any official limits on the right of nationals to live and raise their families there. Puerto Ayora in particular grew with tourism during the 1980s, to approximately 5,000 people; Puerto Baquerizo Moreno kept pace with a threefold increase in its fishing fleet and the opening of a major airport on San Cristóbal. The problem was not so much the tourists themselves, a Darwin Station officer explained in 1982, but what they required. "Tourists are easy to control," he noted. "But more tourism necessitates greater infrastructure, including an expanded local population, which politically is impossible to restrict, and that's what does the damage."[41]

Sounding distinctly like Julian Huxley, National Park director Miguel Cifuentes had asserted in 1978 that "humans can live in the Galápagos, but they must do it in a boldly different way—without pollution, without de-spoliation, without any of the dreadful mistakes of the past." Of course, it did not happen that way: within a decade there was such a building boom along the shores of Academy Bay that one Darwin Station visitor guide compared Puerto Ayora to a frontier boom town.[42] No longer in charge of the park, Cifuentes complained that "mainland Ecuadorians are coming into Puerto Ayora on a daily basis, without jobs, without family, and without resources. I think we already reached our population limit in this town, and still we have no immigration policy."[43]

Underemployed Ecuadorians moved to the Galápagos in search of better jobs, but economic conditions there were worse than on the mainland for people who could not speak the tourists' languages. "What they find—if anything—is manual labor, meager pay," the Darwin Station guide explained. Their arrival produced a sprawl of dense settlement on Santa Cruz and San Cristóbal. Where once Darwin Station biologists could lecture every student in Puerto Ayora's one small school about wildlife conservation and make a profound local impact, by 1990 there were ten overcrowded schools on Santa Cruz and a dozen more on other islands. Station personnel could hardly reach them all. Conservation education for some island students consisted of little more than passing by the obtrusive wildlife murals that adorn local schools. "Galápagos, an ecological paradise, let us protect it with pride," read the caption of one such mural at a downtown Puerto Ayora elementary school. Another depicted Galápagos animals extolling a boy in blue jeans: "The settlers protect us." By the 1990s, however, up to half of the islands' residents came from the mainland, where they had not received even this limited introduction to islands' unique wildlife. An urban lifestyle had largely replaced farming and fishing, and few of the new residents even realized that they lived in an ecological paradise.[44]

In any case, many Galápagos settlers had a very different view of paradise than the ecological one envisioned for them by UNESCO.

"Ecuadoreans are moving here in droves with dreams of striking gold," a United Nations publication warned in 1992. "With few economic opportunities elsewhere in Ecuador, conservation concerns have done little to slow growth."[45] The resident population doubled again during the 1990s—approaching 20,000, with over half of these people concentrated in Puerto Ayora. They sent to congress Eduardo Véliz, a populist champion of increased local control over development in the national park and fishing in the marine reserve. Seven dozen tortoises were found hanging from trees on Isabela Island in 1994—allegedly killed by Véliz's supporters to underscore their demands for local autonomy. Surveying a populist election rally in Puerto Ayora three years later, a Galápagos businessman observed, "Among them there is not a single soul who believes in evolution or natural selection or any of the things that make the Galápagos famous."[46]

UNESCO had envisioned that a farsighted commitment to economic development through eco-tourism would foster local concern for protecting native species. Instead, purely recreational tourism gained a foothold. Puerto Ayora became a popular destination for offbeat surfers and scuba divers, many of whom had no interest in nature cruises. "In the Galápagos Islands, for example, our group of frivolous wave-riders . . . doubtless seemed ridiculous to the earnest Sierra Club types standing on the wild islands," a feature writer for *Surfer* commented in 1997. "Those well-meaning tourists surely asked themselves, on one of the great eco-tourism pilgrimages of all time, blessed with more intellectual raw data than perhaps anywhere on Earth: why are these clowns just doing the same bullshit they do at home? . . . But, by and large, we'd remained concerned only with our mission: surf." Shown some of the laid-back local native animals by a guide, the surfer conceded, "It's like Paradise before the Fall," but quickly added, "honestly, we still just wanted waves." They found their paradise in 10-foot-plus breakers. The islands now host major international surfing tournaments.[47]

Other forms of recreational tourism draw visitors as well. Puerto Ayora serves as a port of call for yachters from Europe and North America heading across the Pacific Ocean. Galápagos water sports, hiking and mountain climbing attract adventure-seeking athletes from around the world. Hotels, restaurants, bars and shops line Puerto Ayora's Charles Darwin Avenue, and wealthy Ecuadorians have begun constructing vacation homes on the islands. International telephone service, cable television and internet cafés break the isolation for tourists and settlers alike. Feature articles on the islands appeared during the 1990s in such once-unlikely places as *Runner's World*, *House & Garden* and *Walking Magazine*. A Galápagos beach even provided the setting for *Sports Illustrated*'s 1998 swimsuit spread, whose sole reference to science came in text superimposed over a reclining supermodel's bare leg: "Charles Darwin's observations of animals here laid the foundation for his theory of evolution." Sexual selection apparently played a feature role in this version of the theory. By decade's end, well over 60,000 tourists were descending on the Galápagos annually. Visit "evolution's lab while it lasts," advised *Business Week* in 1996. "Hedge your bets. Go soon."[48]

More than anything in recent years, the 1992 discovery of commercial quantities of sea cucumbers in Galápagos waters divided local settlers from international conservationists. Some Asians value the sluglike animals as aphrodisiacs and offer premium prices for them. Ecuadorian sea cucumber fishermen, known locally as *pepineros*, rushed into the area even though they could not legally catch the creatures there. Attracted by ten times the profit of their traditional catch, some local fishermen joined in pursuit. Ecuadorian officials at first did little to enforce fishing laws in the Galápagos marine reserve and, under pressure from Véliz, opened the region to sea cucumber fishing for a three-month trial period. After the *pepineros* quickly exceeded their authorized quota and the international conservation community arose in protest, the government abruptly closed the harvest early. Violent demonstrations ensued. Chanting "*Mate Solitario Jorge!*" ("Kill Lonesome George!"), armed *pepineros* twice occupied the Darwin Station during 1995 and threatened to butcher its penned animals before the government agreed to negotiate a new charter for the islands.[49]

Conservationists around the world closely monitored these developments, with UNESCO ultimately threatening to revoke the national park's World Heritage Site status if conditions worsened. "Throughout the confrontation, illegal fishing for sea cucumbers continued almost unabated. The flow of illicit cash allowed other social ills, including drugs and prostitution, to take root on the islands," the conservation magazine *International Wildlife* reported, in a grim article titled "Last Look at Paradise?"[50] Ultimately the government brought illegal fishing somewhat under its control, but failed to restore the sense of innocence that once characterized island life. "We never used to lock our doors," one settler commented in 1999. "Now we buy guard dogs and string up barbed wire."[51]

A travel writer for *Vogue* happened to be on a Galápagos nature cruise during the insurrection. "Armed with the latest Darwinism cribbed from books, I began seeing every incident as a plot point in evolution's open-ended Passion Play," he wrote. "Even the *pepineros'* disturbing conduct can be seen as textbook evolutionary behavior [in which] individuals act for their own short-term benefit, or for that of close relatives.[52]

No one had predicted that the archipelago might become a laboratory for cutthroat Social Darwinism. Certainly not Julian Huxley, who dreamed that the Galápagos would help lead all peoples into a new millennium of enlightened evolutionary humanism. In his 1985 novel *Galápagos*, Kurt Vonnegut envisioned a few survivors, isolated in the archipelago on "the Nature Cruise of the Century," evolving into a more peaceable race of humans after everyone else died of war, famine and disease. By January 1, 2000, however, when thousands of residents and visitors poured into Puerto Ayora's streets to greet what they hailed as a new millennium, observers generally conceded that the outside world had transformed Galápagos society much more than the other way around. Using photographs of newly paved streets, mounting garbage and New Year's Eve revelers burning an Ecuadorian official in effigy, *Natural Geographic* now showed the ugly side of island life: "Galápagos, Paradise in Peril," read the title.[53] Illustrating 21st Century realities, 2001 opened with a grounded oil tanker fouling Galápagos waters with fuel intended for tour boats and fishing vessels.

Conservationists agree that introduced species presently constitute the single gravest threat to the archipelago's unique ecosystem. "Isolation is an island group's fundamental property, strongly influencing the ecology and evolution of species composing the ecosystem," explains German biologist Fritz Trillmich, the leading expert on the endangered Galápagos fur seal. Owing to its isolation, the archipelago has fewer species than a similarly sized area of mainland, but many (if not most) are unique. Mass tourism and immigration now disrupt this age-old solitude. Three passenger jets fly to the islands from Ecuador every day, and large cargo ships dock there a dozen or more times each month. "Tons and tons and tons of food and construction material come to the Galápagos every year," the Darwin Station's Howard Snell notes. "Somewhere in those packages are the introduced species of tomorrow. It's not purposeful nowadays, but it's every bit as damaging as if it was."[54]

Feral goats, dogs and other mammals receive most of the media attention, but scientists worry more about introduced insects and plants. Biologists estimate that hundreds of these foreign species now arrive each year, and some of them spread. Cottony cushion scale first appeared on San Cristóbal in 1982, for example, and now attacks dozens of native plants on ten different islands. "In the absence of its natural predators, it's just gone haywire," Darwin Station entomologist Charlotte Causton states. Aphids, wasps and fire ants have spread even further and faster. As conservationists struggle to contain one type of introduced pest, dozens more arrive amid cargo and travelers. Cruise ships and other interisland watercraft inevitably carry them from island to island. "We are fighting very hard against introduced organisms which are arriving probably every day without us knowing it," National Park director Arturo Izurieta observed in 1994. Galápagos Tortoise researcher Thomas H. Fritts added at the time, "We are constantly at threat of reaching that precipice of irretrievable damage to the island ecosystem."[55] The precipice still looms.

Despite the undeniable scientific and cultural significance of keeping the Galápagos ecosystem intact, crossing the precipice of irreparable damage to it would not immediately and utterly destroy the islands. Hawaii, the other great Pacific archipelago, slid down that slope long ago, yet remains a biological treasure. Over half of Hawaii's native species are lost, but it takes a trained eye to notice the difference within its well-managed national parks and on its undeveloped outer islands. Biologists continue to find new species there. The same could follow for the Galápagos. Many endemic species will survive the scourge of alien wasps, fire ants and cottony cushion scale. Giant tortoises endured worse assaults from sailors and collectors during the past two centuries, yet with the Darwin Station's help their numbers have begun to rebound. Geneticists even talk of cloning Lonesome George. The lava fields of Fernandina should withstand the invasion of most foreign species, preserving niches for its unique native plants and animals. Perhaps, with vigilant protection, even the flightless cormorant and Galápagos penguin will hang on. The archipelago's ecosystem has proved surprisingly resilient in the past.[56]

Taking a long-range view, Galápagos environmental law advisor David A. Westbrook warns that "it is difficult to base sound policy on perceptions of crisis." Better to manage known problems in a consistent fashion than to intervene in a radical but unsustainable manner. "To make a slogan out of it," he writes, "crisis management is usually bad management." The Ecuadorian government took a gradualist approach in its so-called New Law for the Galápagos, which passed in 1998. The New Law grew out of the government's promises during the 1995 insurrection to negotiate revised policies for the islands. Insurrectionist leader Eduardo Véliz originally proposed opening the national park and marine reserve to increased human use. In 1997, however, after Véliz was forced into exile along with the populist Ecuadorian president who backed him, conservationists gained the upper hand in the government. An amendment to the national constitution now authorizes restrictions on domestic immigration to the Galápagos for the first time. The New Law also expands the marine reserve, imposes an inspection and quarantine system on incoming cargo and creates a centralized legal regime for addressing development issues.[57]

The New Law's actual impact will depend on how national officials and local settlers implement it. Conservationists applaud the measure even as they doubt whether any mere law can protect the Galápagos from the consequences of its lost isolation. Sure enough, violence erupted anew by local fishermen against conservationists in 2000, after first the sea cucumber season and then the lobster season closed early due to overfishing. The fishermen had participated in setting catch limits under the New Law, but this did not stop them from protesting when, owing to sharp increases in their ranks, they quickly exceeded their quotas. Enticed by inflated prices for shark fin on the East Asian exotic-seafood market, they also demanded an open season for catching the islands' protected, passive sharks. Twice fishermen attacked national park and Darwin Station facilities on Isabela Island, and repeatedly they threatened conservation officials throughout the archipelago. Intimidated by the protestors, or worse, government leaders acceded to many of the fishermen's demands—leaving conservationists more frustrated than ever. "The take home message is that participatory management schemes that are so cleverly touted world wide do not work in these type of situations," Heidi Snell concludes. "The fishers were happy to attend the meetings especially when they demanded and were given payment to attend them, but in the end they do what they want and the rest of the world be damned!" Longtime Galápagos nature photographer Tui De Roy adds, "The model of conservation which for 40 years has been analagous to the word 'Galápagos' is losing its meaning as we sit here wondering what to do." The leader of Isabela's fishermen counters, "This is our home. We know best what has to be done to preserve the environment and nobody has the right to . . . run us out of business."[58]

None of the looming threats have spoiled the Galápagos as a mecca for scientists and eco-tourists, and as a place of profound wonder. "Years ago, someone calculated that the Station had received over 500 visiting scientists but goodness knows what the total is now," longtime Darwin Foundation officer G. T. Corley Smith noted in 1987. The annual number approached 100 scientists in twenty-two groups by 1999—higher than ever.[59] In 2000, over eighty different vessels ranging in size from six-person sailboats to 100-passenger cruise ships offered tours through the national park. "There is nothing like having a baby sea lion burp in your face," began the first Galápagos travel article at the dawn of a new millennium.[60] Documentaries featuring Galápagos wildlife premiere yearly on television and in science museums around the world, including "The Dragons of Galápagos" in 1998, the 1999 "Voyage to the Galápagos," and the 3-D IMAX extravaganza *Galápagos* shown during 2000 at the Smithsonian Institution.

The archipelago remains "a perpetual source of new things" for scientists more than one and one-half centuries after Darwin first proclaimed it so.[61] Smithsonian researcher Carole Baldwin made her first visit to the islands for the IMAX documentary. "Most biologists dream of visiting the Galápagos; Baldwin is no exception," the program noted. In addition to breathtaking wildlife footage, the documentary featured a descent to the Galápagos ocean floor in a revolutionary submersible research vessel. In thirty deep dives during the expedition, Baldwin and her colleagues collected a dozen previously unknown vertebrate species. "While a scientist may describe an equivalent number of new insects in the rain forest in such a short period of time, doing so with vertebrate species at the tail end of the 20th century is unheard of," a Smithsonian publicist boasted.[62] Hype aside, such finds are not peculiar to Galápagos waters. Rather, Galápagos waters attract oceanographers, who find new things wherever they look. In was the same in 1977, when scientists aboard the deep-sea submersible *Alvin* studying hydrothermal vents along the Galápagos Rift discovered complex communities of utterly unknown animals—giant clams, strange mussels, elongated scale and tube worms—living on geothermal rather than solar energy. Subsequently, oceanographers found similar communities at other volcanically active sites of sea-floor spreading.[63]

Peter Grant, Rosemary Grant and their students stand at the other extreme of Galápagos researchers—returning time and again for decades. They too continually find new things. For example, the Grants' increasing recognition of crossbreeding among Darwin's finches suggests that hybridization plays a larger role in evolution than was previously thought. Galápagos ground finches do breed across species, the Grants find, and their hybrid offspring can thrive where conditions favor intermediate types. "The discovery of superior hybrid fitness over several years suggests that the three study populations of Darwin's finches are fusing," they note, "and calls into question their designation as species." Under strict phylogenetic or breeding-population definitions of species, Peter Grant concludes, only six separate species of Darwin's finches may exist, not the traditional fourteen. Owing to these findings, made more than half a century after David Lack's landmark study, Ernst Mayr once again wonders whether Darwin's finches offer a textbook case of adaptive radiation. The ground finches may indeed constitute a "hybrid swarm" with marked varieties, much as Darwin feared and pre-Lack ornithologists surmised. As such, they could hold even more clues to how evolution works.[64]

David J. Anderson, one the Grants' earliest field-workers, has also made a career of studying Galápagos birds—in his case, the masked boobies of Española Island. He has documented a regular pattern of stronger chicks driving their siblings from the nest in a struggle for survival that begins at birth. "Life stinks for most organisms most of the time, and we maybe don't realize it so much looking at continental systems as opposed to here, because here the animals do their suffering out in the open," Anderson remarks. "We can . . . participate in the community as a kind of ignored observer."[65]

DARWINIAN SCIENTISTS SUCH AS Anderson can see as much horror in the Galápagos as Melville ever did; the creation-minded Agassiz family could find as much innocence there as the evolutionist Annie Dillard. Ever since Bishop Berlanga discovered it in 1535, the archipelago has influenced how people impose meaning on nature and draw both scientific and spiritual insight from the world around them—and it has grown more influential over time. In recognition of its cultural significance, the National Geographic Society chose the Galápagos as one of "fifty places of a lifetime" for twenty-first century travelers and one of the world's ten greatest wild places. "Its importance for travelers and scientists alike stems mainly from its role in the history of human thought," Stephen Jay Gould wrote in justification of the society's choice. Its fearless and conspicuous animals have inspired countless visitors. "The Galápagos fauna therefore becomes an appropriate symbol for a necessary alliance of art and nature in our quest for human understanding, our need to nurture respect for nature's bounty and our inextricable role as one little twig on life's luxuriant evolutionary bush," Gould concluded. Seconding the society's choice, WWF chief scientist Eric Dinerstein added, "The Galápagos are sacred, a cradle of modern thought. Just think of the luminescent ideas that came from such a primitive place."[66]

For the British playwright Tom Stoppard, the harsh and violent terrain cast an awful pall over the place: he quoted the *Beagle*'s Fitzroy in describing it as "a shore fit for pandemonium," the capital of Milton's hell. Yet the animals revealed to Stoppard a glimpse of Eden. "This is more strange than words can make it," he wrote. "One walks among iguanas, herons, doves, mocking-birds, and finches as Adam and Eve in medieval paintings walk among antelopes and cranes. The sea-lion lies down with the snorkler. Boobies nest on the trails."[67] Actress Janet Suzman read these lines at the WWF's gala twentieth-anniversary celebration. Stoppard's biblical imagery thus served the cause of science, communicating the WWF plea for global conservation to a wide audience.[68]

These tensions between paradise and purgatory that all but dumbfounded Stoppard infuse the archipelago with meaning for many reflective observers who find themselves caught twixt life and death, science and religion. Even Melville ultimately saw two sides to the place. "Sackcloth and

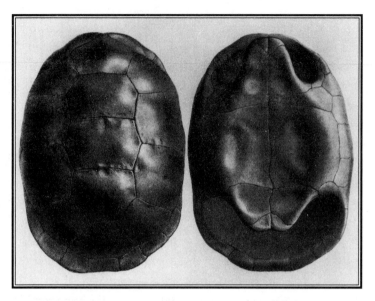

Two sides of a tortoise, from Albert Günther's treatise (1875)
(Source: Used by permission of UGA.)

ashes as they are, the isles are not perhaps unmitigated gloom," he conceded in one sketch. "Even the tortoise, dark and melancholy as it is upon the back, still possesses a bright side; its calapee or breast-plate being sometimes of a faint yellowish or golden tinge." After urging observers to look on this bright side too, Melville warned, "But after you have done this, and because you have done this, you should not swear that the tortoise has no dark side."[69] Melville is remembered in the name of a cross street that runs to the public dock in the Galápagos capital of Puerto Baquerizo Moreno. Fittingly, it intersects at the water's edge with Charles Darwin Avenue.

A century after Melville, David Lack wrote of evolutionary life coming from individual death among Darwin's finches—and the Grants saw it happen on Daphne Major: light and dark merge in a Darwinian worldview. For some, the Galápagos is solely a place for scientific discovery; for others, spiritual meaning lies in the science from this place. "Humans need their stories—grand, compelling stories—that help to orient us in our lives and in the cosmos," biologist Ursala Goodenough writes. "The Epic of Evolution is such a story, beautifully suited to anchor our search for planetary consensus, telling us our nature, our place, our context." The Genesis account of Eden is another epic story that clings to the cultural consciousness. Lack characterized Genesis as "poetic imagery" and maintained that devout Christians like himself could treat it so "without any loss in the spiritual meaning of the story."[70] As long as the Galápagos Islands continue to inspire these images and symbolize such epic stories, they will remain classic ground in human experience.

APPENDIX

ENGLISH NAME	ECUADORIAN NAME
Abingdon	Pinta
Albemarle	Isabela
Barrington	Santa Fé
Bindloe	Marchena
Charles	La Floriana, Floreana or Santa María
Chatham	San Cristóbal
Crossman	Los Hermanos
Culpepper	Darwin
Duncan	Pinzón
Hood	Española
Indefatigable	Santa Cruz
James	Santiago or San Salvador
Jervis	Rábida
Narborough	Fernandina
North Seymour	Seymour
South Seymour	Baltra
Tower	Genoesa
Werman	Wolf

NOTES

Introduction

1. Julian Huxley, "Charles Darwin: Galápagos and After," in Robert I. Bowman, ed., *The Galápagos: Proceedings of the Symposia of the Galápagos International Scientific Project* (Berkeley: University of California Press, 1966), 3. See, e.g., Barbara Sleeper and Michael Konecny, "Darwin's Eden: The Galápagos Wonderland," *Animals*, 120 (September–October 1987): 10.

2. Herman Melville, "The Encantadas, or Enchanted Isles," in Harrison Hayford et al., eds., *The Piazza Tales and Other Prose Pieces: 1839–1860* (Evanston, IL: Northwestern University Press, 1987), 125, 129, 136, 127–28.

3. Charles Roberts Anderson, *Melville in the South Seas* (New York: Columbia University Press, 1939), 33, 113, 431; Jay Leyda, ed., *The Complete Short Stories of Herman Melville* (New York: Random House, 1949), 455–59; Jay Leyda, *The Melville Log: A Documentary Life of Herman Melville*, vol. 1 (New York: Harcourt Brace, 1951), 122–23; I. Newbery, "'The Encantadas': Melville's *Inferno*," *American Literature*, 38 (1966): 778–79; Victor Wolfgang von Hagen, Introduction, in Herman Melville, *The Encantadas, or Enchanted Isles* (Burlingame, CA: Wreden, 1940), x.

4. In 1940, Galápagos travel writer Victor von Hagen wrote of Melville's sketches, "they are the finest descriptive pieces of writing concerning this volcanic archipelago. . . . I know of nothing that describes the islands as poetically and, I am almost tempted to say, as realistically." von Hagen, Introduction, ix.

5. Melville, "Encantadas," 126–27.

6. Ibid., 127.

7. Ibid., 139–40.

8. Ibid., 139, 171, 126.

9. Ibid., 127–28. As to their longevity, no one knows how long giant land tortoises live because it exceeds the time elapsed since the mid-1800s, when people began monitoring them closely. The best clue we have is that the English navigator and explorer Captain

James Cook gave one to the Queen of Tonga in the 1770s, and it finally died of old age in 1966; yet we do not know how old it was when Cook presented it.

10. Ibid., 134–36.

11. Ibid., 136.

12. Ibid., 126.

13. Ibid., 138, 141–42.

14. Ronald L. Numbers, *Creation by Natural Law: LaPlace's Nebular Hypothesis Comes to America* (Seattle: University of Washington Press, 1967).

15. Melville, "Encantadas," 128–29.

16. Just as Melville characterized his whaling ship as his Yale and Harvard, Darwin wrote, "The voyage of the *Beagle* has been by far the most important event in my life, and has determined my whole career." Charles Darwin, *The Autobiography of Charles Darwin*, Nora Barlow, ed. (London: Collins, 1958), 76.

17. Anderson, *Melville*, 50; H. Bruce Franklin, "The Island Worlds of Darwin and Melville," *Centennial Review*, 11 (1967): 354, 363; Merton M. Sealts, Jr., *Melville's Reading*, rev. ed. (Columbia: University of South Carolina Press, 1988), 171.

18. Melville, "Encantadas," 143.

19. Ibid., 130, 134–35.

20. Ibid., 140.

21. Mark Dunphy, "Melville's Turning of the Darwinian Table in 'The Encantadas,'" *Melville Society Extracts*, 79 (1989): 14. See also Franklin, "Island Worlds," 365; Benjamin Lease, "Two Sides of the Tortoise: Darwin and Melville in the Pacific," *The Personalist*, 49 (1968): 533–34.

22. Melville, "Encantadas," 127–32.

23. Ibid. See generally L. D. Gottlieb, "The Uses of Place: Darwin and Melville on the Galápagos," *Bioscience*, 25 (1975): 172–75.

24. "The Valley of the Amazon," *Putnam's Monthly Magazine*, 3 (1854): 272–79.

25. "The Great Cemetery," *Putnam's Monthly Magazine*, 3 (1854): 249–50.

26. In these culture wars, Melville (although highly eclectic in his own beliefs) scorned the new scientific solutions. See, e.g., Herman Melville, *Clarel: A Poem and Pilgrimage in the Holy Land* (New York: Hendricks House, 1960), 522–23. Here Melville uses an unnamed island clearly drawn from his Galápagos memories as a setting for a dialogue between science and religion largely critical of the former.

27. Robert I. Bowman, "Contributions to Science from the Galápagos," in R. Perry, ed., *Key Environments: Galápagos* (Oxford: Pergamon Press, 1984), 278.

28. Julian Huxley, Preface, in Irenäus Eibl-Eibesfeldt, *Survey on the Galápagos Islands* (Paris: Unesco, 1959), 5.

Chapter 1

1. Christopher Columbus, *Journal of the First Voyage, 1492* (Warminster, UK: Aris & Phillips, 1990), 31, 41–43, 49, 51. Although scholars question the word-by-word authenticity of Columbus's *Journal* because the original is lost, most accept it as a fair transcription or abstract dating from very early in the colonizing process. See, e.g., Peter Hulme, *Colonial Encounters: Europe and the Native Caribbean, 1492–1797* (London: Methuen, 1986), 17–18.

2. Ibid., 73, 119; Antonello Gerbi, *Nature in the New World: From Christopher Columbus to Gonzalo Fernández de Oviedo* (Pittsburgh, PA: University of Pittsburgh Press, 1975), 21.

3. Peter Harrison, *The Bible, Protestantism, and the Rise of Natural Science* (Cambridge, UK: Cambridge University Press, 1998), 9, 22, 82, 91.

4. Gerbi, *Nature in the New World*, 86, 114, 278–79.

5. Ibid., 37; Amerigo Vespucci, "This New World," in Victor Wolfgang von Hagen, ed., *South America: The Green World of the Naturalists* (London: Eyre & Spottiswoode, 1951), 15.

6. Gonzalo Fernández de Oviedo y Valdés, "Selections from the Natural History," in von Hagen, ed., *South America*, 22–25. Of Oviedo's "Natural History," Antonello Gerbi writes, "Every animal is considered in and for itself, as an absolute type" and not in relationship to other types. "The sole point of reference is usually geographical . . . to underline in what way the animal of the Indies is like or unlike its European counterpart." Gerbi, *Nature in the New World*, 295–305 (quote on 305).

7. José de Acosta, "Of the Animals of the New World," in von Hagen, ed., *South America*, 57, 68–70.

8. Ernst Mayr, *The Growth of Biological Thought: Diversity, Evolution, and Inheritance* (Cambridge, MA: Harvard University Press, 1982), 309.

9. Victor Wolfgang von Hagen, Introduction, in von Hagen, ed., *South America*, xi. See generally Gerbi, *Nature in the New World*; Konrad Burdach, *Reformation, Renaissance, Humanismus* (Berlin: Gebrüder Paetel, 1918).

10. Gerbi, *Nature in the New World*, 117–23; Peter T. Bradley, *The Lure of Peru: Maritime Intrusion into the South Sea, 1598–1701* (London: Macmillan, 1989), 1–5; von Hagen, Introduction, xiii.

11. Joseph Richard Slevin, "The Galápagos Islands: A History of Their Exploration," *Occasional Papers of the California Academy of Science*, 25 (1959), 11–13; Victor Wolfgang von Hagen, *Ecuador and the Galápagos Islands* (Norman: University of Oklahoma Press, 1949), 177–78.

12. Fray Thomas de Berlanga to Imperial Catholic Majesty, 26 April 1535, reprinted in Slevin, "Galápagos Islands," 14; von Hagen, *Ecuador*, 171.

13. Berlanga to Majesty, 14–16.

14. Ibid., 15.

15. Ibid., 15–16.

16. Columbus, *Journal*, 31.

17. von Hagen, *Ecuador*, 175.

18. John Hickman, *The Enchanted Islands: The Galápagos Discovered* (Dover, NH: Tanger Books, 1985), 21–24; von Hagen, *Ecuador*, 176.

19. Hickman, *Enchanted Islands*, 21.

20. Ibid., 22.

21. Slevin, "Galápagos Islands," 18.

22. Hickman, *Enchanted Islands*, 22.

23. Peter Gerhard, *Pirates of the Pacific, 1575–1742* (Lincoln: University of Nebraska Press, 1960), 13.

24. von Hagen, *Ecuador*, 178.

25. Christopher Lloyd, *William Dampier* (London: Faber & Faber, 1966), 42–44 (quotes buccaneer William Cowley's description of the Danish prize as a "lovely ship"); Gerhard, *Pirates of the Pacific*, 154–56.

26. One friendly biographer writes, "Dampier's story of his first circumnavigation, in the years 1679 to 1691, chronicled the first English achievement of the sort since Cavendish, nearly a hundred years earlier. The fact that so many books of buccaneering exploration followed so close upon the heels of Dampier's books, and were so often written by comrades, associates or rivals of his, such as Wafer, Davis, Backwell, Funnell, Cowley, and Woodes Rogers, tends to obscure the fact that Dampier's books were the earliest and by far the best. He had no worthy successor in geography, in scientific observation or in writing skill, until the arrival of Captain Cook and his 'suite of scientists' on the South Seas scene half a century later." Joseph C. Shipman, *William Dampier: Seaman-Scientist* (Lawrence: University of Kansas Libraries, 1962), 29–30.

27. Ibid., 2–3.

28. William Dampier, *A New Voyage Round the World* (London: James Knapton, 1697), 100–1.

29. Charles Webster, *The Great Instauration: Science, Medicine and Reform* (New York: Holmes & Meier, 1975), 88.

30. Ibid., 17.

31. Ibid., 161, 420; Harrison, *Bible, Protestantism and Science*, 167–68.

32. Lloyd, *William Dampier*, 12–13; Shipman, *Dampier*, 4–5.

33. Dampier, *New Voyage*, A2–3.

34. Lloyd, *William Dampier*, 160 (emphasis added).

35. Dampier mentioned by name only two other Galápagos plants, Burton-wood bushes, which he called "very good firing," and Maummee-trees on "some of the Westermost of these Islands." All three of the land plants grew near the coast. Dampier, *New Voyage*, 101, 295, 311.

36. Ibid., 90.

37. Ibid., 101–2, 109.

38. Ibid., 106.

39. Ibid., 107–10. Other buccaneers who published accounts of this layover on the Galápagos made greater mention of birds on the Galápagos Islands than Dampier. See, e.g., William Cowley, *Voyage Round the Globe*, 3d ed., in *A Collection of Voyages in Four Volumes*, vol. 4 (London: Knapton, 1729), 10; Lionel Wafer, *A New Voyage and Description of the Isthmus of America*, in ibid, vol. 4, 115.

40. See, e.g., Jacques Brosse, *Great Voyages of Discovery: Circumnavigators and Scientists, 1764–1843* (New York: Facts on File, 1983), 13.

41. Cowley, *Voyage Round the Globe*, 10–11; Wafer, *New Voyage and Description*, 115; von Hagen, *Ecuador*, 182–83.

42. von Hagen, *Ecuador*, 181. On the history of Galápagos island names see John M. Woram, "Galápagos Island Names," *Noticias de Galápagos*, 48 (December 1989): 22–32.

43. von Hagen, *Ecuador*, 183. See also Dampier, *New Voyage*, 109–10; James Burney, *A Chronological History of the Voyages and Discoveries in the South Seas or Pacific Ocean*, vol. 4 (London: Luke Hansard & Sons, 1816), 190 (Dampier excerpt). For comment by later sailors and commentators about the impact of these reports see Burney, *Chronological History*, vol. 4, 146–47; Hickman, *Enchanted Islands*, 43.

44. Woodes Rogers, *A Cruising Voyage Round the World* (London: A. Bell, 1727), 211.

45. See generally Lloyd, *William Dampier*, 145–52.

46. According to Dampier's biographer, Rogers's published journal of this expedition "was as successful as Dampier's first book" about the earlier visit on the *Batchelor's Delight*. Ibid., 156.

47. Rogers, *Cruising Voyage*, 207.

48. Edward Cooke, *A Voyage to the South Sea* (London: H.M., 1712), 148. Among the buccaneers on this voyage, Cooke and Rogers were the first to publish narratives about it. "With two such accounts on the market," Dampier's biographer noted, "there was no room and probably no inclination on Dampier's part to add to them." Lloyd, *William Dampier*, 156.

49. Rogers, *Cruising Voyage*, 265.

50. Ibid., 262, 265; Cooke, *Voyage to the South Seas*, 148.

51. Francis Bacon, "New Atlantis," in *The Works of Francis Bacon*, vol. 1 (Philadelphia: Murphy, 1887), 261.

52. John Donne, "To the Countesse of Huntingdon," ll. 1–2, in John Donne, *The Complete English Poems* (New York: Knopf, 1991), 311.

53. Antonello Gerbi, *The Dispute of the New World: The History of a Polemic, 1750–1900* (Pittsburgh, PA: University of Pittsburgh Press, 1973), 5–7.

Chapter 2

1. William Paley, *Natural Theology: Or, Evidence of the Existence and Attributes of the Deity, Collected from the Appearance of Nature* (Philadelphia: John F. Watson, 1814 rpt.), 5–6.

2. For an introductory discussion of this issue see Charles C. Gillispie, *Genesis and Geology: A Study in the Relations of Scientific Thought, Natural Theology, and Social Opinion in Great Britain, 1790–1850* (Cambridge, MA: Harvard University Press, 1951), 7.

3. Numerous historical studies analyze the impact of Paley's natural theology on Darwin's thinking, and cite Darwin's self-analysis on this topic. For a sample of the modern scholarship see Janet Browne, *Charles Darwin: Voyaging* (Princeton, NJ: Princeton University Press, 1995), 97, 129, 529, 542–43.

4. John Ray, Preface, in Charles E. Raven, *John Ray Naturalist: His Life and Work* (Cambridge: Cambridge University Press, 1942), 251.

5. Francis Bacon, *Advancement of Learning and New Atlantis*, Arthur Johnson, ed. (Oxford: Clarendon Press, 1974), 94.

6. John Ray, *The Wisdom of God Manifested in the Works of Creation* (London: Samuel Smith, 1691), xiv, 60–65, 135–50 (quotes on xiv, 63).

7. Idem, in John C. Greene, *The Death of Adam: Evolution and Its Impact on Western Thought* (Ames: Iowa State University Press, 1959), xii.

8. D. L. LeMahieu, *The Mind of William Paley: A Philosopher and His Age* (Lincoln: University of Nebraska Press, 1976), 31.

9. Gillispie, *Genesis and Geology*, 96.

10. For the classic discussion of this trait of science see Thomas S. Kuhn, *The Structure of Scientific Revolutions*, 2d ed. (Chicago: University of Chicago Press, 1962), 52–65.

11. Jacques Brusse, *Great Voyages of Discovery: Circumnavigators and Scientists, 1764–1843* (New York: Facts on File, 1983), 19.

12. George Vancouver, *A Voyage of Discovery to the North Pacific Ocean and Round the World, 1791–1795*, vol. 4 (London: Hakluyt Society, 1984), 193, 1459–66.

13. Herman Melville, *Moby-Dick or the White Whale* (New York: Signet, 1961), 318.

14. James Colnett, *A Voyage to the South Atlantic and Round Cape Horn into the Pacific Ocean* (London: W. Bennett, 1798), 146–48.

15. Ibid., 156, 157–58, 59; Joseph Richard Slevin, "The Galápagos Islands: A History of Their Exploration," *Occasional Papers of the California Academy of Sciences*, 25 (1959): 42.

16. Colnett, *Voyage to the South Atlantic*, 53, 58.

17. Ibid., 52–54.

18. Ibid., 143; Slevin, "Galápagos Islands," 44.

19. Colnett, *Voyage to the South Atlantic*, 48–56, 157.

20. Ibid., 57, 159.

21. On the history of Galápagos island names see John M. Woram, "Galápagos Island Names," *Noticias de Galápagos*, 48 (December 1989): 22–32. According to Woram, the island originally called Charles by the buccaneer William Cowley was misidentified by Colnett, leading him to rename Cowley's Charles as Chatham (now San Cristóbal) and assign the name Charles to the large southern island now popularly known as Floreana.

22. Victor Wolfgang von Hagen, *Ecuador and the Galápagos Islands* (Norman: University of Oklahoma Press, 1949), 199.

23. Slevin, "Galápagos Islands," 39. A typical criticism of errors in Colnett's Galápagos chart, which carries added credibility because it came from a British navy officer, appears in Robert C. Allan, "Remarks on Some of the Galápagos Islands," *Nautical Magazine*, (February 1836): 68.

24. von Hagen, *Ecuador*, 199.

25. Amasa Delano, *A Narrative of Voyages and Travels in the Northern and Southern Hemispheres* (Boston: E. G. House, 1817), 369.

26. Ibid., 379, 376–77.

27. Ibid., 380–81.

28. Ibid., 370–71, 382–83.

29. Harold H. Scudder, "Melville's Benito Cereno and Captain Delano's Voyages," *Publication of the Modern Language Association of America*, 43 (1928): 502; Russell Thomas, "Melville's Use of Some Sources in The Encantadas," *American Literature*, 3 (1932): 432–34.

30. See, e.g., Herman Melville, "The Encantadas, or Enchanted Isles," in Harrison Hayford et al., eds., *The Piazza Tales and Other Prose Pieces: 1839–1860* (Evanston, IL: Northwestern University Press, 1987), 143.

31. David Porter, *Journal of a Cruise* (Annapolis, MD: Naval Institute Press, 1986), 155 (reprint of 1815 ed., noting changes of 1822 ed.).

32. Ibid., 247, 268.

33. Roy Porter, *The Making of Geology: Earth Science in Britain, 1660–1815* (Cambridge: Cambridge University Press, 1977), 184–215.

34. See, e.g., Peter Bowler, *Evolution: The History of an Idea* (Berkeley: University of California Press, 1984), 41–44 (Hutton quote on 44).

35. Porter, *Journal of a Cruise*, 167, 204–5, 232.

36. Ibid., 263, 167, 255, 263. These four escaped goats were the first of many feral animals to flourish on James and other islands, with devastating impact on native species.

37. See, e.g., Bowler, *Evolution*, 106–12.

38. Porter, *Journal of a Cruise*, 255.

39. Ibid., 176, 244.

40. Ibid., 104, 190.

41. "Porter kept up a running dialogue with Colnett in his journal, and used the old captain as a whipping boy for his own observations of the islands." R. D. Madison, Introduction, in Porter, *Journal of a Cruise*, xvii.

42. Benjamin Morrell, *A Narrative of Four Voyages to the South Seas* (New York: J. & J. Harper, 1832), 124–26 (quote on 221). The large, central island that Morrell called Indefatigable originally appeared as a small island named Norfolk on early buccaneer charts and Colnett's map. Perhaps due to sighting a different part of the island and therefore not connecting it with Colnett's Norfolk, later British sailors called it Indefatigable in honor of British Admiral Horatio Nelson's great warship, while many Americans called it Porter in honor of their naval hero. In his journal of his cruising with Porter on the *Essex*, Midshipman William W. Feltus noted, "Porter's isle is not laid down on any map that I have yet seen." Slevin, "Galápagos Islands," 70. The island appeared in somewhat diminutive form as "Porters I." on a chart added to the second (1822) edition of Porter's published journal. Porter, *Journal of a Cruise*, 178.

43. See, e.g., J. N. Reynolds, *Voyage of the United States Frigate Potomac* (New York: Harper, 1835), 465; "Gallipagos Islands—No. 2: Notes of a Terrapin Hunt," *The Friend*, 5 (15 May 1847): 73–74; George Little, *Life on the Ocean* (Baltimore: Armstrong & Berry, 1843), 69–72. See also accounts quoted in von Hagen, *Ecuador*, 207; Slevin, "Galápagos Is-

lands," 56. For a summary account see Bruce C. Epler, "Whalers, Whales, and Tortoise," *Oceans*, 30 (1987): 86–92.

44. Reynolds, *Voyage of the Potomac*, 19–21, 464.

45. Ibid., 464–72 (quote on 467). See also Slevin, "Galápagos Islands," 105–6.

46. "Donations," *Boston Journal of Natural History*, 1 (1834–37): 521; *American Naturalists* (1889): 1039.

47. Bowler, *Evolution*, 61. Linnaeus's category of order is equivalent to the present-day category of family, with a higher category called order added to the system.

48. Vancouver, *Voyage of Discovery*, 1442, 1462–63, 1466. In the world of science, botanist Duncan Porter noted, "Menzies is rarely given credit for having been on the archipelago." D. M. Porter, "The Vascular Plants of Joseph Dalton Hooker's *An Enumeration of the Plants of the Galápagos Archipelago; with Descriptions of Those Which Are New*," *Botanical Journal of the Linnean Society*, 81 (1980): 81.

49. See excepts from the unpublished logs of both ships in Slevin, "Galápagos Islands," 73–74.

50. John Shillibeer, *A Narrative of The Briton's Voyage to Pitcairn's Island* (Taunton, England: J. W. Marriot, 1817), 32.

51. Basil Hall, *Extracts from a Journal, Written on the Coasts of Chili, Peru, and Mexico, in the Years 1820, 1821, 1822* (Edinburgh: Archibald Constable, 1824), 140.

52. Ansel F. Hemenway, "Editorial Prefatory Notes," *Quarterly of the Oregon Historical Society*, 5 (1904): 216–17; W. J. Hooker, "A Brief Memoir of the Life of Mr. David Douglas," *Quarterly of the Oregon Historical Society*, 5 (1904): 228.

53. David Douglas, "Sketch of a Journey to the Northwestern Parts of the Continent of North America During the Years 1824–25–26–27," *Quarterly of the Oregon Historical Society*, 5 (1904): 238–39. In this passage, Douglas writes of the loss of his specimens to rain: "The Gallipagos have been so little visited by scientific persons, that everything becomes of interest which is brought from thence, and I have now little or nothing to show that I have been there!" See also excerpts from Douglas's diary in Ira L. Wiggins and Duncan M. Porter, *Flora of the Galápagos Islands* (Stanford, CA: Stanford University Press, 1971), 37; and a revised edition of Douglas's diary published as David Douglas, *Journal Kept by David Douglas During His Travels in North America, 1823–1827* (New York: Antiquarian Press, 1959 rpt.) (Galápagos Islands portion in brief on 55).

54. Blodwen Lloyd, "John Scouler, M.D., Ll.D., F.L.S. (1804–1871)," *Glasgow Naturalist*, 18 (1962): 210.

55. John Scouler, "Journal of a Voyage to N. W. America," *Quarterly of the Oregon Historical Society*, 6 (1905): 70–71. For Scouler's revised version of this passage from his journal

see Mr. [John] Scouler, "Account of a Voyage to Madeira, Brazil, Juan Fernandez, and the Gollapagos Islands," *Edinburgh Journal of Science*, 5 (1826): 210–14.

56. Wiggins and Porter, *Flora*, 38; Porter, "Vascular Plants," 82. For the history of Scouler's collections see Blodwen Lloyd, "The Herbarium of the Royal College of Science and Technology Glasgow," *Glasgow Naturalist*, 18 (1964): 363–65.

57. Scouler, "Journal of a Voyage," 70–75. This passage includes Scouler's call for further study: "The Gallipoges as will be seen by this very incomplete notice of their productions are peculiarly rich in the objects of scientific research. . . . The rich variety of aquatic birds must satisfy the most zealous ornithologist [and] an island that abounds in so many interesting volcanic appearances . . . will always command the attention of the [geologist]."

58. Lord Byron, *Voyage of H.M.S. Blonde to the Sandwich Islands* (London: John Murray, 1827), 91, 93–94.

59. Porter, "Vascular Plants," 83; Wiggins and Porter, *Flora*, 38.

60. Slevin, "Galápagos Islands," 114; Porter, "Vascular Plants," 83–84; John Thomas Howell, "Hugh Cuming's Visit to the Galápagos Islands," *Lloydia*, 4 (1941): 291–92.

61. Paley, *Natural Theology*, 293.

62. Colnett, *Voyage to the South Atlantic*, 53–54; Donald McLennan, extracts of journal reprinted in von Hagen, *Ecuador*, 207; Porter, *Journal of a Cruise*, 167.

Chapter 3

1. Robert Fitzroy, *Narrative of the Surveying Voyages of His Majesty's Ships Adventure and Beagle*, vol. 2 (London: Henry Colburn, 1839), 486.

2. Charles Darwin, *Journal of Researches into the Geology and Natural History of the Various Countries Visited by H.M.S. Beagle* (London: Henry Colburn, 1839), 454.

3. Richard Darwin Keynes, ed., *Charles Darwin's Beagle Diary* (Cambridge: Cambridge University Press, 1988), 352.

4. John Barrow, "Sketch of the Surveying Voyages of His Majesty's Ships Adventure and Beagle, 1825–1836," *Journal of the Royal Geographical Society of London*, 6 (1836): 312.

5. Pringle Stokes, in Janet Browne, *Charles Darwin: Voyaging* (New York: Knopf, 1995), 146.

6. See Browne, *Charles Darwin*, 146–49; Adrian Desmond and James Moore, *Darwin: The Life of a Tormented Evolutionist* (New York: Warner, 1991), 104. About the captain's personality, Darwin later wrote, "Fitz-Roy's character was a singular one, with many very noble features: he was devoted to his duty, generous to a fault, bold, determined, in-

domitably energetic, and an ardent friend to all under his sway. . . . [But] Fitz-Roy's temper was a most unfortunate one. It was shown not only by passion but by fits of long-continued moroseness against those who had offended him." Charles Darwin, *The Autobiography of Charles Darwin*, Nora Barlow, ed. (London: Collins, 1958), 72–73.

7. J. S. Henslow to Charles Darwin, 24 August 1831, in Frederick H. Burkhardt et al., eds., *The Correspondence of Charles Darwin*, vol. 1 (Cambridge: Cambridge University Press, 1985), 128–29.

8. Josiah Wedgwood II to R. W. Darwin, 31 August 1831, in ibid., 134.

9. Charles Darwin to J. S. Henslow, 5 September 1831, in ibid., 142.

10. Charles Darwin to Charles Whitley, 9 September 1831, in ibid., 150.

11. Darwin to Henslow (1831), 142. In another letter about the same time, Darwin wrote, "The only thing that now prevents me finally making up my mind is the want of *certainty* about S[outh] S[eas] Islands . . . Cap Fitz. says I do good by plaguing Cap Beaufort" of the admiralty. Charles Darwin to Susan Darwin, 9 September 1831, in ibid., 146.

12. Keynes, ed., *Darwin's Beagle Diary*, 392.

13. Darwin, *Autobiography*, 100. Reflecting the initial priority of disciplines, Darwin originally titled his account of the voyage *Journal of Researches into the Geology and Natural History of the Various Countries Visited by H.M.S. Beagle*, only to later give natural history top billing in the second edition as the concept of organic evolution grew to dominate his thinking about the trip. The *Journal of Researches* is now popularly known as *Voyage of the Beagle*.

14. Henslow to Darwin (1831), 129.

15. Adam Sedgwick to Charles Darwin, 18 September 1831, in Burkhardt et al., eds., *Correspondence of Darwin*, vol. 1, 157–58.

16. Darwin, *Autobiography*, 101.

17. Idem, *Journal* (1839), 228–29, 237.

18. Charles Lyell, *Principles of Geology*, vol. 2 (London: John Murray, 1832), 18–35.

19. Charles Darwin to Catherine Darwin, 31 May 1835, in Burkhardt et al., eds., *Correspondence of Darwin*, vol. 1, 449. For a helpful chronology of the voyage see Appendix I, "Chronology 1821–36," in ibid., 540–43.

20. Charles Darwin to Caroline Darwin, July 1835, in ibid., 458; Charles Darwin to J. S. Henslow, 12 July 1835, in ibid., 461.

21. Darwin, *Journal* (1839), 453–55; Fitzroy, *Narrative of Surveying Voyages*, 488.

22. Keynes, ed., *Darwin's Beagle Diary*, 354, 357.

23. Darwin, *Journal* (1839), 455–58.

24. Browne, *Charles Darwin*, 297.

25. Keynes, ed., *Darwin's Beagle Diary*, 353.

26. Ibid., 353; Darwin, *Journal* (1839), 465; Fitzroy, *Narrative of Surveying Voyages*, 488, 498 (stated numbers collected); Frank J. Sulloway, "Darwin's Conversion: The *Beagle* Voyage and Its Aftermath," *Journal of the History of Biology*, 15 (1982): 344.

27. Darwin, *Journal* (1839), 456.

28. Keynes, ed., *Darwin's Beagle Diary*, 354.

29. Ibid., 354, 363.

30. Darwin, *Journal* (1839), 460.

31. Ibid., 454.

32. Charles Darwin to J. S. Henslow, 28–29 January 1836, in Burkhardt et al., eds., *Correspondence of Darwin*, vol. 1, 485; Duncan M. Porter, "The Vascular Plants of Joseph Dalton Hooker's *Enumeration of the Plants of the Galápagos Archipelago; with Descriptions of Those Which Are New*," *Botanical Journal of the Linnean Society*, 81 (1980): 84–88; Ira L. Wiggins and Duncan M. Porter, *Flora of the Galápagos Islands* (Stanford, CA: Stanford University Press, 1971), 38; and Charles Darwin, *Journal of Researches into the Natural History and Geology of Countries Visited During the Voyage of H.M.S. Beagle* (New York: Appleton, [1845]), 461 (undated authorized 2d ed.).

33. Darwin, *Journal* (1839), 465, 467–69, 471; Richard Darwin Keynes, ed., *Charles Darwin's Zoology Notes & Specimen Lists from H.M.S. Beagle* (Cambridge: Cambridge University Press, 2000), 293–96; Browne, *Darwin*, 301.

34. Keynes, ed., *Darwin's Zoology Notes*, 300; Darwin, *Journal* (1839), 475–77 (quote about catching birds).

35. Darwin, *Journal* (1839), 476.

36. Keynes, ed., *Darwin's Beagle Diary*, 359.

37. Lyell, *Principles of Geology*, vol. 2, 64–65.

38. Keynes, ed., *Darwin's Beagle Diary*, 356.

39. Lyell, *Principles of Geology*, vol. 2, 124.

40. Keynes, ed., *Darwin's Beagle Diary*, 356.

41. Lyell, *Principles of Geology*, vol. 2, 90.

42. Darwin, *Journal* (1839), 460. In his personal notebooks compiled during the late 1830s, Darwin repeatedly speculated about how this one Galápagos land mammal (which he consistently called a "Galápagos mouse") had rafted to the Galápagos. See, e.g., Paul H. Barrett et al., eds., *Charles Darwin's Notebooks, 1836–1844* (Ithaca: Cornell University Press, 1987), 226, 248, 354.

43. Lyell, *Principles of Geology*, vol. 2, 103.

44. Keynes, ed., *Darwin's Beagle Diary*, 357; Darwin, *Journal* (1839), 465–66, 472–73.

45. Darwin to Henslow (1836), 485; Nora Barlow, ed., *Charles Darwin and the Voyage of the Beagle* (New York: Philosophical Library, 1946), 247; Darwin, *Journal* (1839), 469.

46. Darwin, *Journal* (1839), 474; Keynes, ed., *Darwin's Zoology Notes*, 291, 298; idem, *Darwin's Beagle Diary*, 360.

47. Nora Barlow, ed., "Charles Darwin and the Galápagos Islands," *Nature*, 136 (1935): 391. Historian Frank J. Sulloway presents a slightly different version of this passage in Sulloway, "Darwin's Conversion," 327–28.

48. Barlow, ed., "Darwin and the Galápagos," 391.

49. Sulloway, "Darwin's Conversion," 327–37.

50. Darwin, *Journal* (1839), 474.

51. "Zoological Society," *Morning Harald* (London), 12 June 1837, 5. Similarly worded comments appeared in other London newspapers about the same time. Sulloway, "Darwin's Conversion," 357n46.

52. "January 10, 1837," *Proceedings of the Zoological Society of London*, n.v. (1837), 4.

53. In his Zoology field notes, Darwin had identified the various Galápagos finches as finches, grosbeaks, "Icterus" (or blackbirds) and wrens or warblers. Keynes, ed., *Darwin's Zoology Notes*, 297–99.

54. "May 10th, 1937," *Proceedings of the Zoological Society of London*, n.v. (1837), 49. Similar observations appear in Darwin's field notes. Keynes, ed., *Darwin's Zoology Notes*, 297.

55. "February 28th, 1837," *Proceedings of the Zoological Society of London*, n.v. (1837), 26–27; Frank J. Sulloway, "Darwin and His Finches: The Evolution of a Legend," *Journal of the History of Biology*, 15 (1982): 20–36; idem, "Darwin's Conversion," 359–69.

56. Gavin de Beer, ed., "Darwin's Journal," *Bulletin of British Museum (Natural History) Historical Series*, 2, no. 1 (1959): 7.

57. Sandra Herbert, ed., *The Red Notebook of Charles Darwin* (Ithaca: Cornell University Press, 1980), 127–28. In his next notebook, Darwin added about Galápagos plants and animals in relation to those of South America, "The type would be of the continent though species all different." Barrett et al., eds., *Darwin's Notebooks*, 173.

58. Charles Darwin, *The Foundations of the Origin of Species: Two Essays Written in 1842 and 1844*, Francis Darwin, ed. (Cambridge: Cambridge University Press, 1909), 33, 180, 187.

59. Sulloway, "Darwin's Conversion," 336–69; Barrett et al., eds., *Darwin's Notebooks*, 482 (exclamation about tortoises); Fitzroy, *Narrative of Surveying Voyages*, 505; Charles Darwin to J. S. Henslow, 28 March 1837, in Burkhardt et al., eds., *Correspondence of Darwin*, vol. 2, 13 (request to Henslow).

60. Darwin, *Foundations*, 159–60.

61. Ibid., 182. Darwin developed this argument (that evolution more rationally explained the similarity of Galápagos species to American ones than special creation) during 1837 and 1838. Thus he jotted in his transmutation notebooks late in 1837, "The question if creative power acted at Galápagos it so acted that Bi[r]ds with plumage & tone of voice partly American." Returning to this similarity a few months later, he characterized the "fact" that "mocking thrushes of Galápagos having tone of voice like S. American" as a "singular coincidence if distinct creation." About the "flora of Galápagos," he added in 1838, "Did Creator make all new yet forms like neighbouring Continent. This fact speaks volumes. . . . My theory explains this but no other will." Also in 1838, he noted, "From these views we can deduce why small islands should possess many peculiar species. . . . Hence the Galápagos Isl[an]ds are explained. On distinct Creation, how anomalous, that the smallest[,] newest, & most wretched isl[an]d should possess species to themselves.—Probably no case in world like Galápagos." Barrett et al., eds., *Darwin's Notebooks*, 195, 296, 305, 640; see also 405, 425.

62. Ibid., 182.

63. Of the inevitable balance between human population and food supply, Malthus wrote, "Nature, in the attainment of her great purposes, seems always to seize upon the weakest part." T. R. Malthus, *An Essay on the Principle of Population* (London: Ward, 1890), 465–71 (quote on 467).

64. Browne, *Charles Darwin*, 387–88. Darwin described this two-stage process of discovery in Charles Darwin, *The Variation of Animals and Plants Under Domestication*, vol. 1, 2d ed. (New York: Appleton, 1897), 9–10.

65. Barrett et al., eds., *Darwin's Notebooks*, 375–76.

66. Browne, *Charles Darwin*, 543. For further analysis of Malthus's impact on Darwin's thinking see Peter Bowler, "Malthus, Darwin, and the Concept of Struggle," *Journal of the History of Ideas*, 37 (1976): 631–50; David Kohn, "Theories to Work By: Rejected Theories, Reproduction, and Darwin's Path to Natural Selection," *Studies in the History of Biology*, 4 (1980): 67–170; Desmond and Moore, *Darwin*, 264–68.

67. Charles Darwin, *Geological Observations on the Volcanic Islands, Visited During the Voyage of H.M.S. Beagle* (London: Smith, Elder, 1844), 114.

68. Darwin, *Journal* (1839), 454, 461–62.

69. See, e.g., Charles Darwin to J. S. Henslow, November 1839, in Burkhardt et al., eds., *Correspondence of Darwin*, vol. 2, 237; Charles Darwin to J. S. Henslow, 22 January 1843, in ibid., 348; Charles Darwin to J. D. Hooker, 28 April 1845, in ibid., vol. 3, 182; Charles Darwin to J. D. Hooker, 11–12 July 1845, in ibid., 216.

70. J. D. Hooker to Charles Darwin, [12 July 1845], in ibid., vol. 3, 221–22.

71. See, e.g., Charles Darwin to J. D. Hooker, [22 July–19 August 1845], in ibid., 226–27.

72. Darwin, *Journal* (1845), 372–73, 390–92. Darwin biographers Adrian Desmond and James Moore describe this passage about modified finches as "a broad clue, and as much as he would ever say on finch evolution." Desmond and Moore, *Darwin*, 328.

73. Charles Darwin to J. D. Hooker, [8 or 15 July 1846], in Burkhardt et al., eds., *Correspondence of Darwin*, vol. 3, 327.

74. See, e.g., Darwin, *Autobiography*, 118; idem, *Variation of Animals*, vol. 2, 9–10.

75. Idem, *Journal* (1845), 388.

76. Desmond and Moore, *Darwin*, xv (Darwin quote).

77. Charles Darwin to J. D. Hooker, May 1846, in Burkhardt et al., eds., *Correspondence of Darwin*, vol. 3, 317. This expedition aboard HMS *Herald* (accompanied by HMS *Pandora*) took six years and, like the *Beagle* expedition, circumnavigated the globe. Regarding Darwin befriending Cuming see, e.g., Hugh Cuming to Charles Darwin, 28 July 1845, in Burkhardt et al., eds., *Correspondence of Darwin*, vol. 3, 230–31.

78. See, e.g., Charles Darwin to J. D. Hooker, 3 September 1846, in ibid. 340; J. D. Hooker to Charles Darwin, 28 September 1846, in ibid., 342; Porter, "Hooker's *Enumeration*," 89–90. About Edmonstone's work with the expedition generally see Berthold Seemann, *Narrative of the Voyage of H.M.S. Herald* (London: Reeve, 1853), 54. On the recommendation of W. J. Hooker, Seemann joined the expedition as naturalist after Edmonstone's death. For portions of the *Narrative* relating to the period prior to his joining the expedition, which included the visit to the Galápagos Islands, Seemann relied on writings and words from other members of the expedition.

79. John Gould, *The Zoology of the Voyage of H.M.S. Beagle: Part III, Birds* (London: Smith, Elder, 1841), 64 (Gould quote, perhaps written by Darwin, who worked closely with Gould on this volume); J. D. Hooker to Charles Darwin, [28 April 1845], in Burkhardt et al., eds., *Correspondence of Darwin*, vol. 3, 183; Charles Darwin, *On the Origin of Species by Means of Natural Selection* (London: John Murray, 1859), 48 (notes dispute over dividing various Galápagos birds into varieties and species).

80. Edward Belcher, *Narrative of a Voyage Round the World, Performed in Her Majesty's Ship Sulphur, During the Years 1836–1842* (London: Henry Colburn, 1843), xix–xxii, 192.

81. The French expeditions involved the vessels *L'Astrolabe*, *La Vénus* and *La Génie*; the Swedish expedition traveled aboard the frigate *Eugenie*. Regarding these expeditions in the Galápagos see Abel du Petit-Thouars, *Voyage Autour du Monde sur la Frégate la Vénus* (Paris: Gide, Editeur, 1840), 282–313, 322; Joseph Richard Slevin, "The Galápa-

gos Islands: A History of Their Exploration," *Occasional Papers of the California Academy of Sciences*, 25 (1959): 93, 96 (reprinted Galápagos portion of *La Génie* journal in its entirety); Porter, "Hooker's *Enumeration*," 89; Jacques Brosse, *Great Voyages of Discovery: Circumnavigators and Scientists, 1764–1843* (New York: Facts on File, 1983), 24–32, 139–60, 185–94. For a catalogue of birds collected on the Swedish expedition see Carl J. Sundevall, "On Birds from the Galápagos Islands," *Proceedings of the Zoological Society of London*, 39 (1871): 124–26. The dominance of British science over the Galápagos is evidenced by this catalogue appearing in London rather than Stockholm.

82. Of the expeditions that passed through the archipelago after the *Beagle* visit and before publication of the *Origin of Species* gave new meaning to Galápagos science, the Swedish one aboard the *Eugenie* (which spent nine days there in 1852) returned with the most Galápagos specimens, including fifty new species of plants, mostly from the previously unexplored interior of Indefatigable Island. See Wiggins and Porter, *Flora*, 38–39.

83. During the mid-1800s, the United States thrice considered buying the Galápagos from Ecuador, but ultimately pulled back after weighing the probable cost of the archipelago against their conceivable value as a whaling station and source for guano, the only uses that the Americans could think of for it. See, e.g., William R. Manning, ed., *Diplomatic Correspondence of the United States: Inter-American Affairs, 1831–1860* (Washington, DC: Carnegie Endowment, 1935), 302, 333; Victor Wolfgang von Hagen, *Ecuador and the Galápagos Islands* (Norman: University of Oklahoma Press, 1949), 249–50.

84. Fitzroy, *Narrative of Surveying Voyages*, 488.

85. Herman Melville, "The Encantadas, or Enchanted Isles," in Harrison Hayford et al., eds., *The Piazza Tales and Other Prose Pieces, 1839–1860* (Evanston, IL: Northwestern University Press, 1987), 128.

86. Darwin, *Origin of Species*, 390, 391, 398–400.

87. William Paley, *Natural Theology or Evidences of the Existence and Attributes of the Deity, Collected from the Appearances of Nature* (Philadelphia: John F. Watson, 1814), 293.

88. Darwin, *Autobiography*, 59, 87.

89. Idem, *Variation of Animals*, vol. 1, 9–10. Darwin first published this passage in 1868, but probably wrote it between 1856 and 1858, or shortly before he wrote the *Origin of Species*. Duncan M. Porter and Peter W. Graham, eds., *The Portable Darwin* (London: Penguin, 1993), 259 (introductory note); Duncan M. Porter to author, 5 September 2000, in author's files.

Chapter 4

1. For a concise analysis of Darwin's scientific method and comments about it from other scientists at the time see David G. Hull, *Darwin and His Critics* (Chicago: University of Chicago Press, 1973), 16–36.

2. William D. Mathews, *Outline and General Principles of the History of Life*, vol. 1 (Berkeley: University of California Press, 1928), 6.

3. Henry Fairfield Osborn, "The Hereditary Mechanism and the Search for the Unknown Factors in Evolution," *Biological Lectures Delivered at the Marine Biological Laboratory of Wood's Hole*, 4 (1895): 97.

4. Karl Frank, *The Theory of Evolution in the Light of Fact* (London: Kegan, 1913), 172.

5. Charles Darwin to Asa Gray, 11 May 1863, in Frederick H. Burkhardt et al., eds., *The Correspondence of Charles Darwin*, vol. 11 (Cambridge: Cambridge University Press, 1999), 403 (Darwin underlined *Creation* and *Modification* once and *or* twice).

6. J. D. Hooker, "Review of the *Origin of Species*," *Gardeners' Chronicle*, 31 December 1859, 1051.

7. Richard Owen, "Darwin on the Origin of Species," *Edinburgh Review*, 111 (1860): 496; Charles Lyell to Charles Darwin, 28 August 1860, in Burkhardt et al., eds., *Correspondence of Darwin*, vol. 8 (Cambridge: Cambridge University Press, 1993), 336.

8. Owen, "Darwin on Origin of Species," 496; T. H. Huxley to Charles Darwin, 12 September 1868, in T. H. Huxley, *Life and Letters of T. H. Huxley*, vol. 1 (New York: Appleton, 1901), 319.

9. Peter J. Bowler, *Life's Splendid Drama: Evolutionary Biology and the Reconstruction of Life's Ancestry, 1860–1940* (Chicago: University of Chicago Press, 1996), 380.

10. P. L. Sclater, Notice, *Proceedings of the Zoological Society of London*, 37 (1869): 133.

11. See P. L. Sclater and Osbert Salvin, "Characters of New Species of Birds Collected by Dr. Habel in the Galápagos Islands," *Proceedings of the Zoological Society of London*, 38 (1870): 322–23.

12. Osbert Salvin, "On the Avifauna of the Galápagos Archipelago," *Transactions of the Zoological Society of London*, 9 (1877): 462.

13. Ibid., 456–62 (quote on 462).

14. Ibid., 461, 509.

15. Ibid., 455–56 (quote on 456); S. Eardley-Wilmot, ed., *Our Journal in the Pacific by the Officers of H.M.S. Zealous* (London: Longmans, Green, 1873), 16–21 (quote on 20).

16. Eardley-Wilmot, *Our Journal*, 21.

17. See, e.g., ibid. and, for a modern prediction to the same effect of the impact of humans on Galápagos animals,, Larry Rohter, "Where Darwin Mused, Strife Over Ecosystem,"

New York Times, 27 December 2000, A1, A8. Alfred, Lord Tennyson, "In Memoriam," pt. 56, l. 15, in Susan Shatto and Marion Shaw, eds., *Tennyson: In Memoriam* (Oxford: Clarendon Press, 1982), 80.

18. Louis Agassiz, *Essay on Classification*, Edward Lurie, ed. (Cambridge, MA: Harvard University Press, 1962), 44.

19. Edward Lurie, "Jean Louis Rodolphe Agassiz," *Dictionary of Scientific Biography*, vol. 1 (New York: Charles Scribner's Sons, 1970), 72.

20. Agassiz, *Essay on Classification*, 136–37.

21. Louis Agassiz, in Louis Agassiz and Elizabeth Agassiz, *A Journey in Brazil* (Boston: Ticknor & Fields, 1868), 33.

22. Mary P. Winsor, *Reading the Shape of Nature: Comparative Zoology at the Agassiz Museum* (Chicago: University of Chicago Press, 1991), 75.

23. Benjamin Pierce to Louis Agassiz, 18 February 1871, in Elizabeth Cary Agassiz, ed., *Louis Agassiz: His Life and Correspondence*, vol. 2 (Boston: Houghton Mifflin, 1886), 690.

24. Charles Darwin, in Edward Lurie, *Louis Agassiz: A Life in Science* (Chicago: University of Chicago Press, 1960), 372. On page 373 of this biography, Lurie suggests that the elderly Agassiz went on the *Hassler* expedition with an open mind to test Darwin's theory, but he makes this assumption on the basis of one letter from Agassiz, whereas the many writings and comments noted herein and Agassiz's prior history suggests a closed mind. Reaching the same conclusion about this episode, Stephen Jay Gould writes, "But Agassiz did not sail only to test evolution in the abstract. He chose his route as a challenge to Darwin, for he virtually retraced—and by conscious choice—the primary part of the *Beagle's* itinerary. The Galápagos were not a convenient way station but a central part of the plot." Stephen Jay Gould, "Agassiz in the Galápagos," *Natural History*, 90 (1981): 10.

25. Louis Agassiz to Benjamin Pierce, 2 December 1871, in *Annals and Magazine of Natural History*, 9 (1872): 13. See generally "The 'Hassler' Expedition," *Nature*, 6 (1872): 354; Elizabeth Agassiz, "A Cruise Through the Galápagos," *Atlantic Monthly*, 31 (1873): 584; and Louise Hall Tharp, *Adventure Alliance: The Story of the Agassiz Family of Boston* (Boston: Little, Brown, 1959), 219, 222.

26. Agassiz, "Cruise Through Galápagos," 579.

27. Compare Elizabeth Agassiz's observations with those of Darwin in *On the Origin of Species*: "We behold the face of nature bright with gladness, we often see the superabundance of food; we do not see, or we forget, that the birds which are idly singing round us mostly live on insects or seeds, and are thus constantly destroying life; or we forget how largely these songsters, or their eggs, or their nestlings, are destroyed by

birds and beasts of prey; we do not always bear in mind, that though food may be now superabundant, it is not so at all seasons of each recurring year." Charles Darwin, *On the Origin of Species by Means of Natural Selection* (London: John Murray, 1859), 62.

28. Agassiz, "Cruise Through Galápagos," 582–83; Agassiz, ed., *Louis Agassiz*, vol. 2, 760; J. Henry Blake to Joseph R. Slevin, 22 July 1938, in Joseph Slevin Papers, box 3, California Academy of Sciences (CAS) Archives, San Francisco, CA.

29. "'Hassler' Expedition," 353. Although this report in *Nature* did not carry Agassiz's byline (unlike two earlier ones from the same expedition) and was penned by a journalist accompanying the expedition, it clearly bore Agassiz's stamp.

30. Louis Agassiz to Benjamin Pierce, 29 July 1872, in Agassiz, ed., *Louis Agassiz*, vol. 2, 763; "'Hassler' Expedition," 354. Elizabeth Agassiz closed her *Atlantic Monthly* article with the very same rhetorical question about the source of Galápagos life that her husband sent to Pierce, and answered it bluntly: "Either it originated where it is found, or else those changes, by whose subtle, imperceptible alchemy it is argued that all differences of species have been brought about, are much more rapid in their action than has been supposed." Agassiz, "Cruise Through Galápagos," 584.

31. In 1868, the great physicist Lord Kelvin calculated that, based on thermodynamics and known energy sources, the earth could not be more than 100 million years old. Most biologists accepted his estimate and looked for evolutionary mechanisms that could generate current life within this time frame. Although Darwin believed that Kelvin's calculations must be wrong, he too conceded an ever greater role for neo-Lamarckian mechanisms in later editions of *On the Origin of Species* at least in part to accelerate the evolutionary process. Only later did the discovery of radioactivity lead physicists to lengthen their estimates of the earth's age. For an introduction to these debates see Peter J. Bowler, *Evolution: The History of an Idea* (Berkeley: University of California Press, 1984), 130, 194–96, 244.

32. Lurie, *Agassiz*, 376.

33. Agassiz to Pierce, 1872, in Agassiz, ed., *Louis Agassiz*, vol. 2, 763; Agassiz, "Cruise Through Galápagos," 584; "'Hassler' Expedition," 359.

34. On Cope see Ronald L. Numbers, *Darwinism Comes to America* (Cambridge, MA: Harvard University Press, 1998), 24, 32–35, 140–41 (quote on 24).

35. Peter J. Bowler, *The Non-Darwinian Revolution: Reinterpreting a Historical Myth* (Baltimore: Johns Hopkins University Press, 1988), 72. Bowler here refers to professional biologists in the United States and western Europe. The situation differed somewhat in the Roman Catholic regions of southern Europe and Latin America, where the church actively defended creationism, sometimes with government support. This had little im-

pact on science in the Galápagos Islands, however, because British and American naturalists conducted research on the archipelago with scant interference from the islands' nominal Ecuadorian sovereigns. For a wide range of national reactions to Darwinism, with material on southern Europe and Latin America, see Thomas F. Glick, ed., *Comparative Reception of Darwinism* (Austin: University of Texas Press, 1974); and Paul J. Weindling, "Darwinism in Germany, France and Italy," in David Kohn, ed., *The Darwinian Heritage* (Princeton, NJ: Princeton University Press, 1985), 683–730.

36. For an overview of this period see Bowler, *Non-Darwinian Revolution*, 71–104.

37. Numbers, *Darwinism Comes to America*, 28; Peter J. Bowler, *Charles Darwin: The Man and His Influence* (Oxford: Basil Blackwell, 1990), 175.

38. See, e.g., Bowler, *Life's Splendid Drama*, 27–28; Ronald Rainger, *An Agenda for Antiquity: Henry Fairfield Osborn and Vertebrate Paleontology at the American Museum of Natural History, 1890–1935* (Tuscaloosa: University of Alabama Press, 1991), 19–23, 54–66. For a general discussion of the rise of natural history museums in the United States see Joel J. Orosz, *Curators and Culture: The Museum Movement in America, 1740–1870* (Tuscaloosa: University of Alabama Press, 1990).

39. Albert Günther, "Description of the Living and Extinct Races of Gigantic Land-Tortoises," *Proceedings of the Royal Society of London* (1875): 252–53.

40. Ibid., 252.

41. Ibid., 258.

42. See, e.g., Albert Günther, "Description of *Ceratodus*, a Genus of Ganoid Fishes, Recently Discovered in Rivers of Queensland, Australia," *Philosophical Transactions of the Royal Society*, 161 (1871): 560.

43. Reflecting years later on the impact of his early work on tortoises, Günther observed, "Zoologists of my younger years took little interest in them, and many a specimen which is now valued as one of the treasures in a collection had been stowed away among the curiosities of the lumber-room." Further, he noted with "satisfaction," his original monograph led to an explosion of more than fifty scientific papers on the topic. Albert Günther, "The President's Anniversary Address," *Proceedings of the Linnean Society of London*, 110 (1898): 18.

44. Günther, "Description of Tortoises," 258.

45. W. E. Cookson, "Report of a Visit by H.M.S. 'Peterel' to the Galápagos Islands," *Proceedings of the Zoological Society of London*, 44 (1876): 178–79.

46. Albert Günther, "Account of the Zoological Collection Made During the Visit of H.M.S. 'Peterel' to the Galápagos Islands," *Proceedings of the Zoological Society of Lon-*

don, 45 (1877): 64. Despite the added specimens, Günther still refrained from hypothe-
sizing about tortoise origins. Indeed, in addition to the slight of listing Darwin after
Fitzroy when discussing scientific work on the Galápagos, this report made a point of
reporting Cookson's observation that "the birds continue to be as tame as in former
times, especially in Charles and Chatham Island, which have been so long inhabited;
the small birds of all kinds are so tame that they are easily knocked down with a
switch"—precisely the same observation made by Darwin. Of course, Darwin had suggested that selection would breed out this vulnerability, but Günther stressed its persistence. Ibid., 65.

47. A. H. Markham, "A Visit to the Galápagos Islands in 1880," *Proceedings of the Royal
Geographical Society*, 2 (1880): 747–50. The birds alighting on gun barrels were probably
mockingbirds. Günther closely monitored all Galápagos Tortoises brought back to European museums. See, e.g., Henry H. Giglioti to Albert Günther, 7 May 1898, A. Günther Collection, 19/1/1, Natural History Museum (NHM) Archives, London, UK.

48. Osbert Salvin, "Notes on Captain Markham's 'Visit to the Galápagos Islands,'" *Proceedings of the Royal Geographical Society*, 2 (1880): 756. Even a casual visitor could still
find unknown species in well-known places—such as the U.S. Navy surgeon who collected specimens of an unknown species of oystercatcher during a routine call at
Chatham Island in 1884. Robert Ridgway, "Description of a Recently New Oyster-
Catcher from the Galápagos Islands," *Proceedings of the United States National Museum*,
9 (1886): 325–26.

49. Salvin, "Notes on Markham's Visit," 757–58; idem, "A List of Birds Collected by Captain A. H. Markham on the West Coast of America," *Proceedings of the Zoological Society of London*, 51 (1883): 419.

50. Joseph Richard Slevin, "The Galápagos Islands: A History of Their Exploration," *Occasional Papers of the California Academy of Sciences*, 25 (1959): 56, 104; Victor Wolfgang
von Hagen, *Ecuador and the Galápagos Islands* (Norman: University of Oklahoma Press,
1949), 238; Alexander Agassiz, "General Sketch of the Expedition of the 'Albatross,'
from February to May, 1891," *Bulletin of the Museum of Comparative Zoology*, 23 (1892):
57; Theodor Wolf, "Ein Besuch der Galápagos Inseln," *Sammlung von Vortragen
Deutsche Volk*, 1 (1879): 257–300.

51. Alexander Agassiz, "Cactaceae in the Galápagos Islands," *Nature*, 53 (1896): 199.

52. In a larger sense, the *Albatross* expeditions marked the entry of the United States as a
player in oceanographic research generally. Partially in response to the *Hassler*'s efforts,
which in failure showed distinct promise, Britain had funded an oceanographic research

vessel of its own, HMS *Challenger*. Its success during the mid-1870s led other countries
to follow suit with specially equipped vessels of their own. The *Albatross* represented
successful entry of the United States in this field. For further discussion see Bowler,
Life's Splendid Drama, 27–28, 389–90.

53. Robert Ridgway, "List of Species of Middle and South American Birds Not Contained
in the United States National Museum," *Proceedings of the United States National Museum*, 4 (1881): 165.

54. Idem, "Scientific Results of Explorations by the U.S. Fish Commission Steamer Albatross," *Proceedings of the United States National Museum*, 12 (1889): 101.

55. Z. L. Tanner, "Report on the Work of United States Fish Commission Steamer Albatross," *Report of Commissioner of Fish and Fisheries for 1887* (Washington, DC: Government Printing Office, 1890), 405.

56. Ridgway, "Scientific Results," 101–2.

57. Alexander Agassiz to Marshall McDonald, 14 March 1891, in *Bulletin of the Museum of Comparative Zoology*, 21 (1891): 194–95; Alexander Agassiz to Marshall McDonald, April 1981, in G. R. Agassiz, ed., *Letters and Recollections of Alexander Agassiz* (Boston: Houghton Mifflin, 1913), 258; Agassiz, "General Sketch," 61.

58. Harry Harris, "Robert Ridgway," *The Condor*, 30 (1928): 5–8, 15–17, 55; Alexander Wetmore, "Robert Ridgway, 1850–1929," *Biographical Memoirs: National Academy of Sciences*, 15 (1932): 57–67; Ernst Mayr, "Robert Ridgway," *Dictionary of Scientific Biography*, vol. 11 (New York: Charles Scribner's Sons, 1975), 443.

59. "The dominance of [Ridgway's] ideas came to an end when polytypic species were much more broadly conceived," Ernst Mayr notes. Mayr, "Robert Ridgway," 443.

60. Robert Ridgway, "Birds of the Galápagos Islands," *Proceedings of the United States National Museum*, 19 (1896): 467–68.

61. Ibid., 459–61. Given the threats to native species posed by introduced ones, as reported by the *Albatross* expeditions, Ridgway here added his voice to a growing chorus of Galápagos researchers by warning, "If we are to acquire a more exact knowledge of this classic fauna, an effort to do so should be made before it is too late." Ibid., 461.

62. George Baur, "The Galápagos Islands," *Proceedings of the American Antiquarian Society* (Worcester, MA: Charles Hamilton, 1892), 9.

63. Alfred Russel Wallace, *Island Life*, 1881 rpt. (Amherst, NY: Prometheus Books, 1998), 275. The identification of the Galápagos as composed of oceanic, uplift islands was also central to Darwin's thinking about them. In his early notebooks on evolution, for example, Darwin wrote, "The wonderful species of Galápagos, must be owing to these islands, having been purely result of *elevation*,—all modern & wholly volcanic," and "By

my theory in volcanic or rising isl[an]d, there ought to be a good many races and doubtful species; how is this at Canary Arch[ipelago]—it is so at Galápagos." Paul H. Barrett et al., *Charles Darwin's Notebooks, 1836–1844* (Ithaca: Cornell University Press, 1987), 408, 500.

64. Baur, "Galápagos Islands," 7; idem, "On the Origin of the Galápagos Islands," *American Naturalist*, 25 (1891): 307–9.

65. Idem, "Origin of the Galápagos," 318 (quote); idem, "Galápagos Islands," 4.

66. idem, "Galápagos Islands," 4; idem, "Origin of the Galápagos," 318. The Royal Academy of Berlin turned down Baur's grant request after concluding that the cost "would probably not be in accordance with the results obtained." This suggested that the Royal Academy saw little promise in Baur's research solving the problem of the mechanism of evolution. Baur, "Galápagos Islands," 5.

67. Idem, "Geography and Travel," *American Naturalist*, 25 (1891): 902–5; idem, "Origin of the Galápagos," 308–9.

68. W. Botting Hemsley, "The Flora of the Galápagos Islands," *Nature*, 52 (1895): 623; Agassiz, "General Sketch," 71; George Baur, "New Observations on the Origin of the Galápagos Islands," *American Naturalist*, 31 (1897): 673 (relates objections to Baur's theory, including the quote from Wolf).

69. Robert Ridgway, "Birds of the Galápagos Archipelago," *American Naturalist*, 32 (1898): 388; W. Botting Hemsley, "Dr. Baur and the Galápagos," *Nature*, 53 (1895): 78.

70. Miriam Rothschild, *Dear Lord Rothschild: Birds, Butterflies and History* (Philadelphia: Balaban, 1983), 57, 229.

71. Ibid., 197, 205.

72. Walter Rothschild and Ernst Hartert, "A Review of the Ornithology of the Galápagos Islands," *Novitates Zoologicae*, 6 (1899): 85–86.

73. Frank B. Webster, "List of Instructions," rpt. in Rothschild, *Dear Lord Rothschild*, 366–68.

74. Rothschild and Hartert, "Review of Ornithology," 86 (Rothschild quote); C. M. Harris to F. B. Webster, 8 February 1898 (telegram stating numbers), Walter Rothschild Correspondence, box 28, folder 2, NHM Archives. In a letter to Rothschild from the Galápagos, Harris predicted, "I think our collection will be much the largest and best yet taken from here." C. M. Harris to Walter Rothschild, Walter Rothschild Correspondence, 28/2 (Harrdssowitz to Hartert), NHM Archives (Harris reported having collected 29 tortoises and about 2,000 birds to this point and estimated a total take of 50 tortoises and 4,000 to 5,000 birds).

75. Regarding his tendency to split species, Rothschild once complained to Günther, "You know I always say that the younger men at the Nat. Hist. Museum are ridiculous

lumpers . . . why I found lumped by Hampson under one species, 4 species of 3 genera."
Walter Rothschild to Albert Günther, 18 April 1902, A. Günther Collection, 19/1/3,
NHM Archives.

76. Rothschild, *Dear Lord Rothschild*, 138–40, 302–4. Most of Rothschild's other natural history specimens went to the British Museum after his death. As Miriam Rothschild explains in this book, her uncle had offered the curatorship of his bird collection to a
young Ernst Mayr, who had collected for Rothschild in New Guinea during the late
1920s. Instead, beginning in 1932, Mayr served as curator of the former Rothschild collection for the American Museum in New York, where he used the collection to work
out his classic *Systematics and the Origin of Species*.

Chapter 5

1. C. M. Harris to F. B. Webster, in Miriam Rothschild, *Dear Lord Rothschild: Birds, Butterflies and History* (Philadelphia: Balaban, 1983), 188–89.

2. At this juncture, Webster wrote to Rothschild, "I feared that if their attention was at-
tracted to our destination [the Californians] might take it into their heads to go there."
Ibid., 194. After Harris and his party returned, Rothschild wrote to Albert Günther, "It
was very lucky they went last year; in 3 years time there will not be a living giant Land
Tortoise of any kind in the Galápagos Islands, 'What a damnable shame', is it not?"
Walter Rothschild to Albert Günther, 14 March 1898, A. Günther Collection, 19/1/2,
Museum of Natural History (MNH) Archives, London, UK.

3. Webster to Rothschild, in Rothschild, *Dear Lord Rothschild*, 193; Robert Cushman Murphy, *Oceanic Birds of South America: A Study of Species of the Related Coasts and Seas, Including the American Quadrant of Antarctica*, vol. 1 (New York: Macmillan, 1936), 3
(statement about Beck).

4. C. M. Harris to F. B. Webster, 21 June 1897, in Rothschild, *Dear Lord Rothschild*, 196.

5. W. P. Gibbons, "Circular of the California Academy of Natural Science," June 1853, in
California Academy of Sciences History, box 1, file 1, California Academy of Sciences
(CAS) Archives, San Francisco, CA.

6. Theodore Henry Hittell, *The California Academy of Sciences: A Narrative History,
1853–1906*, Alan E. Leviton and Michele L. Aldrich, eds. (San Francisco: California
Academy of Sciences, 1997), 13.

7. See Michael L. Smith, *Pacific Visions: California Scientists and the Environment, 1850–1915*
(New Haven: Yale University Press, 1987), 56–58.

8. See J. D. Whitney, *California Academy of Sciences President's Annual Address, Jan. 6, 1868* (San Francisco: Basqui, 1868), 14–15; Joseph Ewan, "San Francisco as a Mecca for Nineteenth-Century Naturalists," in *A Century of Progress in the Natural Sciences, 1853–1953* (San Francisco: California Academy of Sciences, 1955), 32.

9. See "Louis Agassiz," *San Francisco Mining and Scientific Press*, 1 August 1872, 129 (front page of issue); "Louis Agassiz," *San Francisco Chronicle*, 1 September 1872, 1; "Professor Agassiz," *San Francisco News Letter*, 7 September 1872, 2; "Academy of Science," *San Francisco Chronicle*, 3 September 1872, 3; "Natural History of the Animal Kingdom," *San Francisco Mining and Scientific Press*, 26 October 1872, 262–65.

10. Louis Agassiz, "Address to the Academy," 2 September 1872, in *Proceedings of the California Academy of Sciences*, 4 (1868–72): 254–56; David Starr Jordan, *The Voice of the Scholar* (San Francisco: Elder, 1903), 148.

11. Hittell, *California Academy of Sciences*, 243–45; "Academy of Sciences," *Daily Evening Bulletin* (San Francisco), 6 January 1874, 1; "Academy of Sciences," *San Francisco Chronicle*, 13 July 1889, 5. Commenting on the impact of Agassiz's address, CAS president George Davidson noted, "His stirring appeal on behalf of science and the reception given to him by the Academy aroused the dormant desire for scientific study and imagination." George Davidson, *California Academy of Sciences President's Annual Address, Jan. 6, 1873* (San Francisco: Basqui, 1873), 4–5.

12. Orrin Leslie Elliott, *Stanford University: The First Twenty-Five Years* (New York: Arno Press, 1977), 76–92, 123.

13. Ibid., 148–49; Jordan, *Voice of Scholar*, 63, 148; Charles B. Turrill, "The Early Days of the Academy," in Hittell, *California Academy of Sciences*, 511 (quoting Agassiz's admonition); Robert C. Miller, "A Resume History of the California Academy of Sciences," 7, in California Academy of Sciences History, box 1, file 1, CAS Archives.

14. David Starr Jordan, ed., *Leading American Men of Science* (New York: Holt, 1910), 166–67; Peter C. Allen, *Stanford: From the Foothills to the Bay* (Stanford, CA: Stanford Historical Society, 1980), 27.

15. David Starr Jordan and Vernon Lyman Kellogg, *Evolution and Animal Life* (London: Sidney Appleton, 1907), 49.

16. See Smith, *Pacific Visions*, 57–59.

17. See, e.g., Jordan and Kellogg, *Evolution*, 52–53 (refers to "struggle for existence"). Jordan once wrote to Harvard University president Charles W. Elliott about the evolutionary process, "I think that science teaches the rise of love out of the ruthless ages more clearly than it does the social Darwinism of the Germans." David Starr Jordan to Charles W.

Elliott, May 1919, David Starr Jordan Papers, SC 58, series I-A, 97–869, Stanford University Archives, Stanford, CA.

18. As to scientists and their religion, Ernst Mayr (who professes to be an atheist and is perhaps the most significant evolutionary biologist of the twentieth century) wrote near the end of his long life, "Virtually all scientists known to me personally have religion in the best sense of this word, but scientists do not invoke supernatural causation or divine revelation." Ernst Mayr, *This Is Biology: The Science of the Living World* (Cambridge, MA: Harvard University Press, 1997).

19. See Jordan and Kellogg, *Evolution*, 52–53, 206.

20. Agassiz, LeConte and Jordan were not church members, but they occasionally attended Unitarian services. Each of them wrote extensively about religion, and Jordan's papers at the Stanford University library contain numerous letters and lectures by him about his religious beliefs. Jordan's biographer concluded that Jordan "contended that he was [religious], and perhaps we should take him at his word. He often wrote of true religion and its value in maintaining morality and as a means of preserving a healthy outlook upon life. Yet he had a lofty disdain for most of the great religions of history. . . . He once told a student gathering that if he had not been brought up in the Christian faith he would have chosen to be a follower of Shintoism, which he interpreted as nature worship." Edward McNall Burns, *David Starr Jordan: Prophet of Freedom* (Stanford, CA: Stanford University Press, 1953), 183.

21. David Starr Jordan, "Isolation as a Factor in Organic Evolution," in American Association for the Advancement of Science, *Fifty Years of Darwinism: Modern Aspects of Evolution* (New York: Holt, 1909), 86–91.

22. See, e.g., ibid., 80–83; David Starr Jordan, *Foot-Notes to Evolution: A Series of Popular Addresses on the Evolution of Life* (New York: Appleton, 1905), 12, 192. Jordan participated in identifying fish brought back by the Hopkins Stanford Galápagos Expedition and assisted with other scientific publications resulting from the effort. See Robert E. Snodgrass to David Starr Jordan, 19 June 1903, and Edmund Heller to David Starr Jordan, 1–6 February 1902, David Starr Jordan Papers, SC 58, series I-A, 37–362 and 31–305, Stanford University Archives, Stanford, CA.

23. See, e.g., "A Trip to the Galápagos Islands," *Pioneer or California Monthly Magazine*, 9 (February 1858): 97, 103.

24. See, e.g., George Baur, "On the Origins of the Galápagos Islands," *American Naturalist*, 25 (1891): 217–29, 307–19.

25. For example, "The great interest attributed to the fauna of the Galápagos Islands since Darwin's exploration, became still more intense through the recent explorations of

Messrs. Baur and Adams." Walter Rothschild and Ernst Hartert, "A Review of the Or-nithology of the Galápagos Islands," *Novitates Zoologicae*, 6 (1899): 86. See also Jonathan Weiner, *The Beak of the Finch: A Story of Evolution in Our Time* (New York: Vintage Books, 1995), 36; John van Denburgh, "The Gigantic Land Tortoises of the Galápagos Archipelago," *Proceedings of the California Academy of Sciences*, 4th ser., 2 (1914): 228. Baur had collected more ground finches in the Galápagos than all prior naturalists combined.

26. The phrase came from Rothschild in a letter to Albert Günther about the capture of Galápagos Tortoises. Rothschild, *Dear Lord Rothschild*, 197. See also Walter Rothschild, "Description of a New Species of Gigantic Land Tortoise from Indefatigable Island," *Novitates Zoologicae*, 10 (1903): 119. A similar objective with respect to the CAS expedi-tion was noted in Joseph R. Slevin, "Log of the Schooner 'Academy': On a Voyage of Scientific Research to the Galápagos Islands, 1905–06," *Occasional Papers of the Califor-nia Academy of Sciences*, 17 (1931): 3. The title of Joseph Slevin's personal account of the CAS expedition to the Galápagos also captured this motivation: "Race With Extinc-tion." The drive to collect the last of dying breeds was promoted in press accounts of the CAS expedition. See, e.g., Edward Berwick, "California Scientists Invade the South Seas," *San Francisco Chronicle*, 18 May 1905, 7.

27. On the results of Rothschild's expeditions see Walter Rothschild to Albert Günther, 10 August 1901, and Walter Rothschild to Albert Günther, 26 August 1901, A. Günther Collection, 19/1/3, MNH Archives; Walter Rothschild and Ernst Hartert, "Further Notes on the Fauna of the Galápagos Islands," *Novitates Zoologicae*, 9 (1902): 373. As these sources indicate, Green was accompanied by a man named Johnson, who did most of the tortoise collecting while Green focused on birds. On Beck's activities see his au-tobiographical notes reprinted in Murphy, *Oceanic Birds*, 3–4; R. H. Beck, "In the Home of the Giant Tortoises," in New York Zoological Society, *Seventh Annual Report* (1902), 160–62; R. H. Beck, "Bird Life Among the Galápagos Islands," *The Condor*, 4 (May–June 1902): 5–11; Walter Rothschild to Albert Günther, 12 October 1902, A. Gün-ther Collection, 19/1/3, MNH Archives (reports on Beck's 1902 trip: "25 dead tortoises, 35 living ones, 600 odd birds & a quantity of odds and ends"). On the CAS expedition see van Denburgh, "Gigantic Land Tortoises," 228–42; Berwick, "California Scientists," 7; "Will Spend a Year in Search of Rare Species," *San Francisco Chronicle*, 21 June 1905, 16. At the time, Americans and Englishmen often were called by their initials rather than their given names. This book follows that convention by attempting to identify persons as they typically identified themselves.

28. See "Two California Methodists Who Will Visit South America in Scientific Re-search," *California Christian Advocate*, undated clipping in R. H. Beck Papers, Ida M.

Menzies scrapbook, 74, CAS Archives. Stewart later received his Ph.D. in botany for his flora of the Galápagos.

29. In every instance, these collectors' Galápagos experience featured prominently in their curricula vitae and obituaries. See, e.g., Elwood C. Zimmerman, "Francis Xavier Williams, 1882–1962," *Pan-Pacific Entomologist*, 45 (1969): 137–38.

30. Rothschild, *Dear Lord Rothschild*, 197–98, 228.

31. Rothschild and Hartert, "Further Notes," 373.

32. "A Notable Collecting Trip," *Stanford Alumnus*, 3 (1899): 71. See also "Papers for the Hopkins Stanford Galápagos Expedition, 1898–1899," *Proceedings of the Washington Academy of Sciences*, 3 (1901): 363–64.

33. See "Papers from the Hopkins Stanford Galápagos Expedition, 1898–99," *Proceedings of the Washington Academy of Sciences*, 3–6 (1901–5). To build the overall collection, university curators exchanged some duplicates for items from other museums, including the trade that sent eighteen of Stanford's tortoises to Rothschild. Rothschild and Hartert, "Further Notes," 373; Walter Rothschild to Albert Günther, 10 August 1901, A. Günther Collection, 19/1/3, MNH Archives (gives number as 16).

34. Berwick, "California Scientists," 7. Regarding Loomis's purpose for the CAS expedition see also Slevin, "Log of the 'Academy,'" 3.

35. "Vessel Brings Galápagos Party," *San Francisco Call*, 30 November 1906 (includes quote); Slevin, "Log of the 'Academy,'" 5; Thomas H. Fritts and Patricia R. Fritts, eds., *Race with Extinction: Herpetological Notes of J. R. Slevin's Journey to the Galápagos, 1905–06* (Washington, DC: Herpetologists' League, 1982), 86–87.

36. "Vessel Brings Party."

37. "Under Auspices of Academy of Science," *San Francisco Examiner*, 21 May 1905.

38. Berwick, "California Scientists," 7; Slevin, "Log of the 'Academy,'" 5. Based on his travels to the Galápagos for Rothschild, Beck was convinced that island tortoises and other endemic species would soon die out like "the American bison after the hide hunters began their work of extermination." Beck, "Home of the Giant Tortoises," 161.

39. "Yacht to Be Used by Scientists Christened," *San Francisco Examiner*, 25 June 1905.

40. F. M. Dickie, "The 'Academy,'" in Galápagos Islands Collection, box 11, unnumbered file, CAS Archives.

41. Fritts and Fritts, eds., *Race with Extinction*, 64; Slevin, "Log of the 'Academy,'" 114–18.

42. C. D. Bunker to Frank B. Webster, 11 February 1898, Walther Rothschild Correspondence, 40/8 (Webergurg to Webster), MNH Archives (quote); Walter Rothschild and Ernst Hartert, eds., "Diary of Charles Miller Harris," *Novitates Zoologicae*, 6 (1899): 101, 104.

43. Slevin, "Log of the 'Academy,'" 162.

44. Ibid., 112, 148, 153.

45. J. S. Hunter, "Galápagos Diary," manuscript, in J. S. Hunter Papers, box 1, CAS Archives.

46. Ibid., 42–43; Rothschild and Hartert, eds., "Diary of Harris," 97–98 (quote); idem, "Notes from the Diary of Mr. F. P. Drowne," *Novitates Zoologicae*, 6 (1899): 123–24 (quote).

47. See, e.g., Slevin, "Log of the 'Academy,'" 127, 139–42; Rothschild and Hartert, eds., "Diary of Drowne," 128.

48. See, e.g., Slevin, "Log of the 'Academy,'" 40–41; Rothschild and Hartert, eds., "Diary of Harris," 96; idem, "Diary of Drowne," 121; R. H. Beck, "Field-Notes on the Tortoises of the Galápagos Islands," *Novitates Zoologicae*, 9 (1902): 378.

49. Drowne commented at one point in his diary, "A more barren place could scarcely be imagined." Rothschild and Hartert, eds., "Diary of Drowne," 121.

50. Slevin, "Log of the Academy,'" 49, 159.

51. Rothschild and Hartert, eds., "Diary of Drowne," 131.

52. Idem, "Diary of Harris," 92, 100.

53. Fritts and Fritts, eds., *Race with Extinction*, 5.

54. Slevin, "Log of the 'Academy,'" 149–50.

55. Ibid., 91–92; Fritts and Fritts, eds., *Race with Extinction*, 52–53. Beck, who could not swim, had a similar brush with death when trying to board a small skiff with a large iguana on an earlier expedition to the islands for Rothschild. "Just as I jumped a sudden swirl caught the boat and swept it past. The iguana landed in the boat, but I, with the gun, dropped into the 'raging main.' Fortunately the mate shoved his oar out, and as I came up minus the gun the oar got in my way, and I was soon in the boat." Beck, "Field-Notes on Tortoises," 377.

56. See, e.g., Rothschild and Hartert, "Further Notes," 373–74; Slevin, "Log of the 'Academy,'" 5.

57. Rothschild and Hartert, eds., "Diary of Harris," 94; idem, "Diary of Drowne," 116.

58. Walter Rothschild, "Description of a New Species of Gigantic Land Tortoise from Indefatigable Island," *Novitates Zoologicae*, 10 (1903): 119; Rothschild and Hartert, "Further Notes," 373.

59. Slevin, "Log of the 'Academy,'" 53.

60. Ibid., 48–49.

61. Fritts and Fritts, eds., *Race with Extinction*, 32, 75; Rothschild and Hartert, eds., "Diary of Drowne," 120; [Annotated List of Tortoise Collection], California Academy of Sciences, Herpetology Department, 1999 (reference listing).

62. Fritts and Fritts, eds., *Race with Extinction*, 72.

63. Ibid., 74–76; Slevin, "Log of the 'Academy,'" 135, 143; [Annotated List of Tortoise Collection].

64. "President's Annual Report," 7 January 1907, California Academy of Sciences, *Stated Meetings Minutes, Dec. 1904–Nov. 1914* (corporate minute book), 92; [Annotated List of Tortoise Collection].

65. "Council Meeting," 16 April 1906, California Academy of Sciences, *Council Minutes, July 1907–May 1914* (corporate minute book), 127–28.

66. R. H. Beck field notes, in van Denburgh, "Gigantic Land Tortoises," 302.

67. Michael H. Jackson, *Galápagos: A Natural History* (Calgary, Canada: University of Calgary Press, 1993), 104–5.

68. Rothschild and Hartert, eds., "Diary of Drowne," 114.

69. See, e.g., Rothschild and Hartert, eds., "Diary of Harris," 106–8; Hunter, "Galápagos Diary," 20, 36; Slevin, "Log of the 'Academy,'" 73, 102, 130; and Rothschild and Hartert, "Review of Ornithology," 142 (includes quotes about cormorant).

70. F. X. Williams, untitled manuscript notes, Francis Xavier Williams Papers, box 4, CAS Archives.

71. Rothschild and Hartert, "Review of Ornithology," 86; Rothschild, *Dear Lord Rothschild*, 201; Rothschild and Hartert, "Further Notes," 381–418; "President's Annual Report, 1907," 92–93; Edward Winslow Gifford, "The Birds of the Galápagos Islands," *Proceedings of the California Academy of Sciences*, 4th ser., 2 (1913): 2. The CAS expedition exceeded prior Galápagos expeditions in its other collections as well. For example, all prior expeditions to the archipelago had collected specimens of less that 40 species of insects, while the CAS expedition brought back specimens of over 150 insect species. F. E. Blaisdell, "23rd Regular Quarterly Meeting," *Entomological News*, 18 (1907): 260–61.

72. David Lack, in Rothschild, *Dear Lord Rothschild*, 138; idem, *Darwin's Finches* (Cambridge: Cambridge University Press, 1948), 165.

73. Rothschild and Hartert, "Review of Ornithology," 136 (this is one of Rothschild's most theoretical studies of his Galápagos material, yet it never speculates about the causes of evolution).

74. Ibid., 136.

75. Robert Evans Snodgrass and Edmund Heller, "Papers from the Hopkins Stanford Galápagos Expedition, 1898–1899: Birds," *Proceedings of the Washington Academy of Sciences*, 5 (1904): 275–77. At this point in their article, Snodgrass and Heller cautioned, "The position of the species as given in the diagram certainly represents their degrees of resemblances, but *we do not claim that it certainly represents their natural relationships.*

We have no way of determining to what extent *convergent evolution* has operated in causing forms to resemble one another." Ibid., 275.

76. H. Gadow, "The Wings and the Skeleton of *Platacrocorax Harrisi*," *Novitates Zoologicae*, 9 (1902): 169–70.

77. Edmund Heller, "Papers from the Hopkins Stanford Galápagos Expedition, 1898–99: Reptiles," *Proceedings of the Washington Academy of Sciences*, 5 (1903): 46–47.

78. van Denburgh, "Gigantic Land Tortoises," 366–68.

79. Williams, untitled notes.

80. Rothschild and Hartert, "Review of Ornithology," 136.

Chapter 6

1. After an initial burst of scientific reports appeared during the first decade of the new century (mostly from Stanford zoologists seeking recognition for their new institution), articles describing island animals and plants appeared from researchers at Tring Museum and the CAS over the next quarter century. Regarding the Stanford publications see David Starr Jordan to Timothy Crocker, 7 February 1900, David Starr Jordan Papers, SC58—ser. I-AA, Jordan letter box 1, Stanford University Archives, Palo Alto, CA. In an effort to gain East Coast recognition for Stanford science, Jordan had the initial reports from the Hopkins Stanford Galápagos Expedition published in volumes three through five of the *Proceedings of the Washington Academy of Sciences* (1901–4).

2. See, e.g., William Coleman, *Biology in the Nineteenth Century: Problems of Form, Function, and Transformation* (New York: Wiley, 1971), 160–66; Hamilton Cravens, *The Triumph of Evolution: The Heredity-Environment Controversy, 1900–1941* (Baltimore: Johns Hopkins University Press, 1988), 15–55; Ronald Rainger, *An Agenda for Antiquity: Henry Fairfield Osborn and Vertebrate Paleontology at the American Museum of Natural History, 1890–1935* (Tuscaloosa: University of Alabama Press, 1991), 19–20, 132–34.

3. Ronald Rainger, "Henry Fairfield Osborn," in John A. Garraty and Mark C. Carnes, eds., *American National Biography*, vol. 12 (Oxford: Oxford University Press, 1999), 785–88; "Dr. Henry F. Osborn Dies in His Study," *New York Times*, 7 November 1935, 23. Osborn also took turns leading various local, state or national scientific associations, including the prestigious American Association for the Advancement of Science in 1927.

4. Henry Fairfield Osborn, *Creative Education in School, College, University and Museum: Personal Observation and Experience of the Half-Century 1877–1927* (New York: Scribner's, 1927), 252.

5. Idem, *Evolution and Religion in Education: Polemics of the Fundamentalist Controversy of 1922 to 1926* (New York: Scribner's, 1926), 54. His biographer adds, "Osborn endowed the study of nature with sanctity, and his theory of evolution upheld the traditional convergence of science and religion." Rainger, *Agenda for Antiquity*, 132. For Osborn's battle with fundamentalists see Edward J. Larson, *Summer for the Gods: The Scopes Trial and America's Continuing Debate Over Science and Religion* (New York: Basic Books, 1997), 113–35. For Osborn's battle with Morgan, who Osborn had hired for the Columbia faculty but could not control, see Rainger, *Agenda for Antiquity*, 132–41.

6. Henry Fairfield Osborn, *Fifty-Two Years of Research, Observation and Publication, 1877–1929: A Live Adventure in Breadth and Depth* (New York: Scribner's, 1930), 151.

7. See Rainger, *Agenda for Antiquity*, 121. For those not privileged by personal wealth or scientific training to accompany these expeditions of self-fulfillment, zoo and museum exhibits offered the next best thing. Under Osborn, the Bronx Zoo pioneered techniques to display living animals in their native habitats, and the American Museum portrayed dead ones in lifelike poses and settings. See Osborn, *Creative Education*, 260.

8. David Starr Jordan, "Personal Glimpses of Theodore Roosevelt," *Natural History*, 19 (1919): 15 (written in eulogy). See also Rainger, *Agenda for Antiquity*, 47; H. W. Brands, *T. R.: The Last Romantic* (New York: Basic Books, 1997), 29–34, 65–66.

9. When the former president died in 1919, Osborn led New Yorkers in mourning by erecting at a tributary exhibit hall at the American Museum an installation commemorating "Roosevelt the Naturalist." Henry Fairfield Osborn, "Theodore Roosevelt, Naturalist," *Natural History*, 19 (1919): 10; idem, *Impressions of Great Naturalists* (New York: Scribner's, 1928), 275 (Roosevelt was one of the "great naturalists" commemorated in this book, along with the likes of Darwin and Wallace).

10. Gifford Pinchot, "Roosevelt, the Man of Abundant Life," *Natural History*, 19 (1919): 17.

11. Theodore Roosevelt, *African Game Trails* (New York: Scribner's, 1911), vii, ix.

12. Gifford Pinchot, *To the South Seas* (Philadelphia: Winston, 1930), 5–8.

13. "Roosevelt in Pacific Lands 45-Pound Tuna," *New York Times*, 29 July 1938, 2.

14. Theodore Roosevelt, in Edwin Emerson, *Adventures of Theodore Roosevelt* (New York: Dutton, 1928), vii.

15. For the motto, purpose and nature of this expedition (including reference to the party's "casual collecting") see William Beebe, *Galápagos: World's End* (New York: Putnam, 1924), vii, xii, 194, 429.

16. "Find Giant Tortoise in Galápagos Isles," *New York Times*, 16 May 1923, 9.

17. "H. Williams Dead, Utilities Leader," *New York Times*, 11 November 1953, 31.

18. William Beebe, "Williams Galápagos Expedition," *Zoologica* 5 (1923–25): 3 (quote), 7; Beebe, *Galápagos*, 3–4; "Garden Party Sees Beebe's Wild Birds," *New York Times*, 22 May 1923, 19.

19. Henry Fairfield Osborn, "In the Wake of Darwin," in Beebe, *Galápagos*, viii.

20. Tim M. Berra, *William Beebe: An Annotated Bibliography* (Hamden, CT: Archon, 1977), 44–58, 109–111.

21. Raymond Pearl to Henry Fairfield Osborn, 3 September 1925, Osborn Papers, correspondence files, American Museum of Natural History Archives, New York (hereafter Osborn Papers); Beebe, *Galápagos*, 11, 194–201 (second Rose quote and her reference to Osborn as "the Scientist"); Ruth Rose, "William Beebe's Voyage in the Galápagos," *New York Herald*, 27 May 1923, 3.

22. Berra, *William Beebe*, 59–60. The book *Galápagos: World's End* was largely original and borrowed little from Beebe's articles.

23. Keir B. Sterling, "William Beebe," in Garraty and Carnes, eds., *American National Biography*, vol. 2, 462.

24. Beebe, *Galápagos*, 166–78. Similarly embellished accounts of the Williams expedition's exploits appear throughout *Galápagos: World's End* and in William Beebe, "The Strangest Islands in the Seven Seas," *Travel*, May 1924, 33–40. Beebe's repeated claim to be the first person to capture a flightless cormorant alive resulted in a complaint from Mary Harris McCaullaugh. Her deceased brother, Charles Harris, accomplished this feat twenty-five years earlier for Walter Rothschild. Osborn apologized for the error. Mary H. McCaullaugh to Henry Fairfield Osborn, 2 December 1924, and Henry Fairfield Osborn to Mary H. McCaullaugh, 13 December 1924, Osborn Papers, correspondence files.

25. See "Flightless Birds in Cargo for Zoo," *New York Times*, 17 May 1923, 14. For Beebe's account of the tortoise see Beebe, *Galápagos*, 224–29.

26. Twenty-one scientific articles about the Williams Galápagos Expedition appear in *Zoologica*, 5 (1923–25), the house journal of the Zoological Society of New York.

27. "William Beebe Sets Sail," *The Nation*, 25 February 1925, 203.

28. Beebe, *Galápagos*, 54 (Osborn), 65–66 (humans), 123–25 (iguanas), 165–75 (cormorants), 266 (finches). Beebe's view of iguana evolution also appears in Edmund Wilson, "A Conversation in the Galápagos: Mr. William Beebe and a Marine Iguana," *Atlantic Monthly*, November 1925, 579.

29. Beebe, *Galápagos*, 65–66. For Beebe's view of the place of humans in nature see also Wilson, "Conversation in Galápagos," 577–87.

30. Beebe, *Galápagos*, 96–101, 157–58 (quote). See also Robert Henry Welker, *Natural Man: The Life of William Beebe* (Bloomington: Indiana University Press, 1975), 86. Beebe's benign view of evolution fit the work of earlier evolutionary naturalists such as LeConte, Jordan and Osborn, but it ignored the newer findings of experimental biologists who were then laying the foundation for a revival of Darwinian natural selection.

31. See Larson, *Summer for the Gods*, 7, 31, 113–35.

32. Beebe, *Galápagos*, 65; "The Arcturus's Cargo," *New York Times*, 1 August 1925, 10; "Interest in Science," *New York Times*, 14 January 1926, 24.

33. William Beebe, *The Arcturus Adventure* (New York: Putnam's, 1926), v–vii; Welker, *Natural Man*, 98–99; "Beebe Ship Silent 11 Days as Radio Calls to Her in Vain," *New York Times*, 10 April 1925, 1, 4 (description of ship); "Beebe Sets Sail," 204 (quote); "Boy Is Author at 12," *New York Times*, 23 July 1925, 13 (noting that Putnam went on the expedition with his entire family, and that his son wrote a book about the experience); "3 New Yorkers Marooned on Beebe's Ship," *New York Times*, 23 June 1925, 1. Williams sailed the *Warrior* alongside for part of the trip.

34. "Operator on Board Arcturus Tells of His Experience," *New York Times*, 30 August 1925, sec. 8, 14. For a representative example of front-page coverage see William Beebe, "Beebe Discovers a New Island in the Pacific; Gets Strangest Fish Seen, with Lighting Rod," *New York Times*, 3 May 1925, 1. The supposed "new island" noted in this article was an abovewater area within the bay of a larger Galápagos Island that did not appear on Beebe's charts. Beebe named it for Osborn, but the name did not survive.

35. William Beebe, "Ocean Tells New Tales to Beebe," *New York Times*, 9 August 1925, sec. 4, 1–2; Beebe, *Arcturus Adventure*, 176 (quote); E. W. Gudger, "Origin of Diving Helmet," *New York Times*, 28 May 1925, 20 (letter to editor by American Museum associate correcting Beebe). On the earlier use of the diving helmet see E. W. Gudger, "On the Use of the Diving Helmet in Submarine Biological Work," *Natural History*, 18 (1918): 135–37. From the expedition, Beebe wrote to Osborn about diving, "It gives you such an intimate sense of contact and oneness with that indescribable world which seems so much richer and more varied than the world above the surface." William Beebe to Henry Fairfield Osborn, 22 June 1925, Osborn Papers, correspondence files.

36. William Beebe, "Beebe Climbs a Fiery Volcano," *New York Times*, 21 June 1925, sec. 4, 1–2 (quote about climb); "Beebe Discovers Two New Volcanoes Erupting Magnificently in Galápagos," *New York Times*, 17 April 1925, 1 ("first eruptions" quote and states that Beebe named the twin volcanos for sponsors Williams and Whiton).

37. R. L. Duffus, "'Arcturus', Whither Away?" *New York Times*, 23 May 1926, sec. 3, 1; Theodore Roosevelt, "William Beebe's 'Jungle Peace,'" *New York Times*, 13 October 1918,

sec. 7, 1. Nicholas Roosevelt, a younger cousin of the former president, made similar observations in his review of *Galápagos: World's End* for the *Times*. Nicholas Roosevelt, "By Sea to the End of the World," *New York Times*, 24 February 1924, sec. 3, 1.

38. Henry Fairfield Osborn to Madison Grant, 18 November 1918, in Osborn Papers, correspondence files. For an assessment of Beebe's impact on public interest in the Galápagos see Victor Wolfgang von Hagen, *Ecuador and the Galápagos Islands* (Norman: University of Oklahoma Press, 1949), 251–55.

39. von Hagen, *Ecuador*, 251–56; Revenue Act of 1924, *United States Statutes*, 68 Cong., 1 sess., ch. 234, sec. 214(a)(10) (1924). For an example of the attention to tax matters see Templeton Crocker to California Academy of Sciences, 15 February 1933; and California Academy of Sciences, "Gift Tax Form 719," California Academy of Sciences correspondence file, 1932, A-C, Templeton Crocker Expedition to Galápagos file, in California Academy of Sciences (CAS) Archives, San Francisco, CA.

40. "Saw Volcanic Eruption," *New York Times*, 18 March 1926, 9 (quote); "Vanderbilt Back with Sea Marvels," *New York Times*, 1 April 1926, 1. The trip netted specimens of thirty-one different species of fish from Galápagos waters, two of them new to science. This was not a particularly notable haul from a tropical archipelago. N. A. Borodin, "Scientific Results of the Yacht *Ara* Expedition During the Years 1926 to 1928, While in Command of William K. Vanderbilt," *Bulletin of the Vanderbilt Oceanographic Museum*, 1 (1928): 1–3, 8–9. Vanderbilt described this expedition in his book *To Galápagos on the Ara* (New York: n.p., 1927).

41. Pinchot, *South Seas*, 5 (quote), 8–11; Sidney Nichols Shurcliff, *Jungle Islands: The "Illyria" in the South Seas* (New York: Putnam, 1930), 4. About the Pinchot expedition see also "Seek Huge Sea-Bat on Pinchot Cruise," *New York Times*, 14 March 1929, 12; "The Unexplored Galápagos," *New York Times*, 28 March 1929, 28. Each expedition returned with two live Galápagos Tortoises. Charles Hastings Townsend, "The Astor Expedition," *Bulletin of the New York Zoological Society*, 33 (1930): 142.

42. Suyden Cutting, *The Fire Ox and Other Years* (New York: Scribner's, 1940), 288.

43. "Again the Galápagos," *New York Times*, 25 April 1930, 24. See also "Notes," *Natural History*, 30 (1930): 326–27; "Astor Party Plans Scientific Cruise," *New York Times*, 20 March 1930, 29; "Nourmahal Party Brings Rare Finds," *New York Times*, 24 April 1930, 24; "Astor Yacht Brings Tropical Rarities," *New York Times*, 3 May 1930, 11.

44. Eunice Fuller Bernard, "Palaces That Cruise the Seven Seas," *New York Times*, 20 April 1930, sec. 5, 12. Despite the luxury, this commentator asserted that Galápagos yachting expeditions restored a sense of adventure to tired lives: "The new sport of kings of finance is the treasure hunt at the ends of the earth. And the fact that it is a

zoological, a botanical or an archeological treasure hunt makes it none the less exciting," she wrote. "The Galápagos Islands, off the South American coast, has become the Mecca." Ibid., 13.

45. Wealthy American anglers had institutionalized the sport a quarter century earlier, complete with published rules and records, by founding the Tuna Club on Catalina Island, a California winter resort frequented by Easterners. "Fishing," *New Encyclopedia Britannica*, vol. 4 (Chicago: Encyclopedia Britannica, 1997), 802; Arthur N. Macrath, Jr., *History of the Tuna Club* (Avalon, CA: n.p., 1948), 12–22.

46. George Reiger, *Zane Gray: Outdoorsman* (Englewood Cliffs, NJ: Prentice-Hall, 1972), vii–xi, 344–49 (complete list of Gray's publications about fishing); Ed Zern, Foreword, in Zane Gray, *Adventures in Fishing* (New York: Harper, 1952), xiii–xv. The combined sales for all of his books made Gray the best-selling American author of the early twentieth century.

47. Zane Gray, *Tales of Fishing Virgin Seas* (New York: Harper, 1925), 47 (about Beebe), 50 (quote).

48. Ibid., 66–87; idem, "Fishing Virgin Seas," *Pacific Sportsman*, August 1925, 1.

49. Henry W. Fowler, "The Fishes of the George Vanderbilt South Pacific Expedition, 1937," *Academy of the Natural Sciences of Philadelphia Monographs*, 2 (1938): 1–2, 6–8; "Results of the Fifth George Vanderbilt Expedition (1941)," *Academy of the Natural Sciences of Philadelphia Monographs*, 6 (1944): 1–5, 303–52; Evelyn Trent, "Bay Group Seeks Galápagos Rights," *San Francisco Chronicle*, 21 April 1929, F–9; "Warns on Galápagos Sale," *New York Times*, 27 June 1930, 6; "Galápagos Islands Attract Sportsmen and Naturalists," *New York Times*, 20 July 1930, sec. 8, 12; "Deserted Galopagos [*sic*] Islands Valued for Their Wild Life," *New York Times*, 2 November 1930, sec. 9, 6; "Ecuador Will Control Access to Galápagos," *New York Times*, 5 March 1938, 9 (foreign "fishing boats must obtain permits from consulates in Panama, San Francisco or San Diego"). For an example of earlier U.S. efforts to acquire the Galápagos and Ecuadorian reaction to them see "Says We Want Galápagos," *New York Times*, 22 January 1911, 1; "Hoot Galápagos Lease," *New York Times*, 28 January 1911, 4; "To Drop Galápagos Lease," *New York Times*, 30 January 1911, 5.

50. "Fear of 'Yankee Invasion' of Galápagos Isles Stirs Ecuadoreans to Seek Foreign Leases," *New York Times*, 23 October 1930, 1.

51. von Hagen, *Ecuador*, 252–55; William Albert Robinson, *10,000 Leagues Over the Sea* (New York: Brewer, 1932), 58–60; Cutting, *Fire Ox*, 291–94; Temple Utley, *A Modern Sea-Beggar* (London: Peter Davies, 1938), 112–13, 154–58. Originally called Charles by the

buccanneers, the southernmost inhabited island in the group has been called La Floriana, Floreana and Santa Maria by Ecuadorians.

52. Utley, *Modern Sea-Beggar*, 120–35, 156–57 ("Nietzschean"); Edwin O. Palmer, *Third Galápagos Trip of the* Velero III *in the Winter of 1933–1934* (Los Angeles: n.p., 1934), 14–16 ("neurotic"); William Albert Robinson, *Voyage to the Galápagos* (New York: Harcourt, 1936), 199–216 (individual titles); Hakon Mielche, *Let's See If the World Is Round* (London: William Hodge, 1933), 108–23 (house names); Irving Johnson, *Westward Bound in the Schooner* Yankee (New York: Norton, 1936), 56–71 ("back-to-nature existence"); Edward S. Sullivan, "The Nudist Empress of the Galápagos," *Real Detective*, May 1935, 50–53; "Tragedy Rules 'Garden' of Nudists," *San Francisco Examiner*, 17 March 1936; "California Couple Building Empire on Volcanic Island in Galápagos," *New York Times*, 1 March 1938, 23; von Hagen, *Ecuador*, 255–56; "Menace Ritter's Privacy," *New York Times*, 16 April 1930, 11 (proposed German settlement); "Estonian Settlers," *New York Times*, 29 October 1933, sec. 4, 8. Further offsetting the American presence, a few European scientific expeditions to the South Pacific passed through the Galápagos during this period, most notably, in 1924, the Norwegian Zoological Expedition and a party of English naturalists and travelers aboard the venerable yacht *St. George* collecting for the British Museum. A. J. A. Douglas and P. H. Johnson, *The South Seas of Today: Being an Account of the Cruise of the Yacht* St. George *to the South Pacific* (London: Cassell, 1928), xi–2, 256–72; C. L. Collenvette, *Sea-Girt Jungles: The Experiences of a Naturalist with the "St. George" Expedition* (London: Hutchinson, 1926), xi, 72–116; L. J. Chubb, "The St. George Scientific Expedition," *Geological Magazine*, 62 (1925): 369–73; Ira L. Wiggins and Duncan M. Porter, *Flora of the Galápagos Islands* (Stanford, CA: Stanford University Press, 1971), 40.

53. Timothy Banning, "The Hancock Legacy," *Westways*, February 1979, 27–30, 73–77; "The Allan Hancock Foundation of the University of Southern California," *ASC Newsletter*, February 1980, 1–5; DeWitt Meredith, *G. Allan Hancock: A Pictorial Account of One Man's Score in Fourscore Years* (San Jose, CA: Paramount, 1964), 58; C. McLean Fraser, "General Account of the Scientific Work of the *Velero III* in the Eastern Pacific, 1931–41," *Allan Hancock Pacific Expeditions*, 1 (1943): 9–21.

54. Martha B. Darbyshire, "Aboard the *Velero III*," *Country Life*, August 1936, 31–32; Fraser, "General Account," 21–23, 50–54; George Hugh Banning, "Hancock Expedition of 1933 to the Galapagós [*sic*] Islands," *Bulletin of the Zoological Society of San Diego*, October 1933, 2–3.

55. Mielche, *Let's See*, 105.

56. Darbyshire, "Aboard the *Velero*" 35–36; Banning, "Hancock Expedition," 4–18.

57. Palmer, *Third Galápagos Trip*, 2–3, 43 (quote); Fraser, "General Account," 43 (about crew); Mielche, *Let's See*, 105; J. T. Howell, "Velero Ensemble," Hancock Papers, unnumbered file, CAS Archives (handwritten note and printed performance program).

58. Harry S. Swarth, "The Avifauna of the Galápagos Islands," *Occasional Papers of the California Academy of Sciences*, 18 (1931): 19 (quote), 137–207.

59. Ibid., 20, 27, 29–30.

60. Templeton Crocker, "Introductory Statement," *Proceedings of the California Academy of Sciences*, 4th ser., 21 (1933): 3–5; C. E. Grunsky, Foreword, ibid., 1; California Academy of Sciences, "Joint Meeting of the Trustees and the Council," 19 September 1932, 3, Barton W. Evermann Papers, box 36, CAS Archives; "Ornithological Results of the 'Zaca' Expedition," Barton W. Evermann Papers, file 36, CAS Archives; J. T. Howell, "The 1932 Templeton Crocker Expedition and J. T. Howell," 22 May 1975, 1, Marin County Library Oral History Project, Marin County, CA. "The manner in which this expedition was organized and conducted affords an excellent example of converting what ordinarily might be a pleasant but prosaic yachting trip into a cruise which yielded results of great interest and permanent scientific value," a CAS official maintained during those Depression-era days of shrinking research budgets. "After all, much of the future exploration of the ocean and oceanic islands must necessarily be done by such men who have the means and the interest to carry out similar projects successfully." G. Dallas Hanna, "The Templeton Crocker Expedition to the Galápagos Islands," typescript, 2, California Academy of Sciences correspondence files, 1932, A–C, Templeton Crocker Expedition to Galápagos file, CAS Archives.

61. "Yacht Sails on Tropical Science Trip," *San Francisco News*, 19 March 1932, 6; Howell, "1932 Templeton Crocker Expedition," 3; California Academy of Sciences, "Joint Meeting," 3; "Crocker Party Climbs Peak on Isle," *San Francisco Call-Bulletin*, 10 May 1932, 1; "Crocker Expedition Back from Tropic Cruise," *San Francisco Chronicle*, 2 September 1932, 1. Regarding the 1934 expedition, which did little on the Galápagos, see James P. Chapin, "To Polynesia on the Yacht 'Zaca,'" *Natural History*, 36 (1935): 293. To commemorate the expedition, the summit of Indefatigable Island is named Mt. Crocker.

62. Howell, "1932 Templeton Crocker Expedition," 2.

63. Idem, "Botanizing with the Templeton Crocker Expedition of the California Academy of Sciences," University of California Botanical Seminar, 6 February 1933, 8–9, Galápagos Island Collection, box 9, CAS Archives.

64. See, e.g., ibid, 5–7.

65. Idem, "Environment and Evolutionary Trends Among Some Plants of the Galápagos Islands," American Association for the Advancement of Science lecture, n.d., 7–8, Galápagos Island Collection, box 9, CAS Archives.

66. H. G. Wells, Julian S. Huxley and G. P. Wells, *The Science of Life*, vol. 2 (Garden City, NY: Doubleday, 1931), 617–21.

67. Howell, "Environment," 1.

68. Charles Haskins Townsend, "The Galápagos Islands Revisited," *Bulletin of the New York Zoological Society*, 31 (1928): 148–69.

69. "Urges Use of Galápagos," *New York Times*, 21 March 1933, 15 (about Swarth); P. R. Lowe, "On the Need for the Preservation of Galápagos Fauna," *Proceedings of the Linnean Society of London*, 146 (1934): 85.

70. On the initial decree and the international effort behind it see Robert T. Moore, "The Protection and Conservation of the Zoological Life of the Galápagos Archipelago," *Science*, 82 (1935): 519–21. Copies of both decrees are retained in Galápagos Committee files (British Association for the Advancement of Science [BAAS]), DF 206/162, Museum of Natural History (MNH) Archives, London, UK. Letters in these files credit Swarth's key role in the process. See, e.g., Harold Coolidge to Julian Huxley, 3 June 1935, Galápagos Committee files (BAAS), DF 206/160, MNH Archives. See also Robert I. Bowman, "The Charles Darwin Foundation," *Pacific Discovery*, 24 (1971): 19–20 (credits Swarth and Coolidge).

71. Edward A. Chapin and Clara Cutler Chapin, "Darwin and the Galápagos Islands," *Bulletin of the Pan-American Union*, 64 (1935): 666 (describes law); "Saving the Galápagos," *New York Times*, 13 October 1935, sec. 10, 4 (von Hagen quote); "Galápagos Wild Life," *New York Times*, 14 June 1936, sec. 11, 9; "Victor Wolfgang von Hagen," in Maxine Black, ed., *Current Biography: Who's News and Why, 1942* (New York: Wilson, 1942), 860.

72. See Julian S. Huxley to W. T. Calman, 26 July 1935 (describes Coolidge as "very keen and in touch with everybody's concern" regarding the Galápagos effort); and Julian S. Huxley to Harold Coolidge, 24 July 1935 (discusses effort to fund a game warden), Galápagos Committee files (BAAS), DF 206/160, MNH Archives. After initially lending support to his broader proposal for a research station, Huxley (warned by Coolidge) shunned von Hagen as a self-interested promoter. See, e.g., Julian S. Huxley to C. G. Darwin, 30 May 1936, Galápagos Committee files (BAAS), DF 206/161, MNH Archives (von Hagen "undoubtedly has done certain good things, but on the other hand equally undoubtedly seems to be an adventurer"). In response to an inquiry from Huxley, Coolidge sent two confidential letters to Huxley highly critical of von Hagen.

Harold Coolidge to Julian Huxley, 6 August 1935 ("personal publicity for von Hagen plays an important part in any set-up which he is connected with"); and Harold Coolidge to Julian Huxley, 12 September 1935 ("he has already embarrassed our Government by some of his actions"), Galápagos correspondence files, DF 206/160, MNH Archives. Charles Darwin's son, Leonard, who had originally endorsed von Hagen's efforts, also came to regret it. Leonard Darwin to O. J. R. Howarth, 21 June 1935 ("I feel guilty in having been a little too kind to von Hagen"); and C. G. Darwin to Julian Huxley, 27 May 1936 ("Hagen has been acting as rather a blight on my family for years now"), Galápagos Committee files (BAAS), DF 206/161, in MNH Archives. The BAAS ultimately endorsed Huxley's proposal over von Hagen's. BAAS Galápagos Islands Committee, "Minutes of Meeting," 20 July 1936, Galápagos Committee files (BAAS), DF 206/159, MNH Archives.

73. P. R. Lowe, "The Finches of the Galápagos in Relation to Darwin's Conception of Species," *Ibis*, 78 (1936): 310, 320–21. For other objections made at this meeting to the sufficiency of natural selection to account for evolution on the Galápagos Islands see "Praises Darwin as Observer," *New York Times*, 7 September 1935, 8.

74. Julian Huxley to O. J. R. Horwarth, 21 May 1937, Galápagos Committee files (BAAS), DF 206/160, MNH Archives ("All ornithologists who have investigated the systematics of the family agree that such an investigation will be of extreme importance, and their variation is unique").

75. David Lack, *Darwin's Finches* (Cambridge: Cambridge University Press, 1983 rpt.), 1 (quote); idem, "My Life as an Amateur Ornithologist," *Ibis*, 115 (1973): 427–28.

76. Idem, "My Life," 427 (island quotes); idem, "The Galápagos Finches: A Study in Variation," *Occasional Papers of the California Academy of Sciences*, 21 (1945): 1–2 (bird quote); idem, *Darwin's Finches*, xv–xvii, 10.

77. Ernst Mayr, *Systematics and the Origin of Species* (New York: Columbia University Press, 1942), 155. For the synthesis to work, geneticists had to abandon their focus on individuals as opposed to populations, and naturalists had to relinquish their reliance on anything other than random, inborn genetic variations selected for their survival value. The neo-Darwinian synthesis did not become dogma within science until after the Second World War, by which time Lack's interpretation of evolution among Galápagos finches was well-known to Huxley, Mayr and other key proponents of the new view. In his book *The Structure of Scientific Revolutions*, philosopher of science Thomas Kuhn describes revolutions in science occurring when old theories give way under the collective weight of accumulated anomalous observations and are replaced by new theories that can ac-

count for those anomalies. Although the neo-Darwinian synthesis was more of an evolution than a revolution in modern biological thought, the Galápagos finches represented one conspicuous anomaly under older understandings of evolution that became *the* textbook example from nature of how the new synthesis explains the origin of species. Thomas S. Kuhn, *The Structure of Scientific Revolutions*, 2d ed. (Chicago: University of Chicago Press, 1970), 52–65. Mayr describes the new synthesis as becoming ascendant among evolutionary biologists following a major international symposium on the issue held at Princeton, New Jersey, in 1947. Lack attended that symposium. Ernst Mayr, *The Growth of Biological Thought: Diversity, Evolution, and Inheritance* (Cambridge, MA: Harvard University Press, 1982), 568–69.

78. Lack, "Galápagos Finches," 116–23 (quotes on 120, 122); idem, "Evolution of the Galápagos Finches," *Nature*, 146 (1940): 324–27; Ernst Mayr, interview with author, 18 November 1999, Cambridge, MA.

79. Percy R. Lowe, "Evolution of the Galápagos Finches," *Ibis*, 83 (1941): 315–17; David Lack, "The Evolution of the Galápagos Finches," *Ibis*, 83 (1941): 637–38 (Lack's response).

80. Julian Huxley, *Evolution: The Modern Synthesis* (New York: Harper, 1942), 242–43, 326.

81. Lack, "My Life," 429.

82. G. F. Gause, "Discussion of Paper by Thomas Park," *American Midland Naturalist*, 21 (1939): 255.

83. Lack, *Darwin's Finches*, 62. Although he does not expressly credit him on this point, Lack may have been influenced by British naturalist R. C. L. Perkins, whose writings on Hawaiian birds anticipated Lack's ideas on the role of competition in the evolution of Galápagos finches. Peter R. Grant, "R. C. L. Perkins and Evolutionary Radiations on Islands," *Oikos*, 89 (2000): 198–200.

84. Ibid, 63.

85. Beebe, *Galápagos*, 261, 266. For Lack's commitment to rulebound order in nature see David Lack, *Evolutionary Theory and Christian Belief: The Unresolved Conflict* (London: Methuen, 1961), 81–82.

86. Lack, *Darwin's Finches*, 113. Lowe originally proposed the name *Darwin's Finches* for the *Geospizids* in a 1935 paper read before the British Association for the Advancement of Sciences during the ceremonies commemorating the centenary of the voyage of the *Beagle*, but the name did not stick until Lack used it in his book. See Percy R. Lowe, "The Finches of the Galápagos" 310.

87. Lack, *Darwin's Finches*, 159, 162.

88. Ernst Mayr, "David L. Lack," *Ibis*, 115 (1973): 433.

89. Lack, "My Life," 430.

Chapter 7

1. Victor Wolfgang von Hagen, "The Chronology of the Darwin Memorial Expedition, and the Plans for the Darwin Research Station" (typescript), 1, Vertical File 506, CDRS/ECCD, I–Z, Charles Darwin Research Station Library, Puerto Ayora, Galápagos Islands; idem, *Ecuador and the Galápagos Islands* (Norman: University of Oklahoma Press, 1949), 258–60.

2. Office of Chief of Naval Intelligence (U.S. Navy), *Field Monograph of Galápagos Islands*, O.N.I.78, 2 February 1942, 232–69 (declassified transcript copy in author's file).

3. See, e.g., ibid., xli–xlvi, 8–9.

4. "Beachhead on the Moon," *Time*, 15 July 1946, 45; "Taps at Galápagos," *Newsweek*, 15 July 1946, 57.

5. J. M. Waram, "Who Killed the Iguanas?" *Noticias de Galápagos*, 50 (1991): 12–13.

6. William Beebe, *Galápagos: World's End* (New York: Putnam, 1924), 243–46.

7. Naval Intelligence, *Field Monograph*, 259. The reporter covering the base closure for *Time* magazine did not list iguanas among the many animals that he found on the island. "Beachhead on the Moon," 45.

8. Irenäus Eibl-Eibesfeldt, *Galápagos: The Noah's Ark of the Pacific* (Garden City, NY: Doubleday, 1961), 177.

9. Idem, "Challenge of the Galápagos," *Nature Magazine*, 50 (1957): 439.

10. The time was ripe for the science of ecology. The specter of world war with its global impact and attendant widespread environmental devastations contributed to ecological thinking. The global population explosion added to the concern in that for the first time, in the 1950s, scientists began seriously to worry that the number of people was approaching the earth's carrying capacity. The ongoing technological revolutions in transportation and communications (which were accelerated by the war effort) further spurred global thinking. See Stephen Bocking, *Ecologists and Environmental Politics: A History of Contemporary Ecology* (New Haven: Yale University Press, 1997), 9–37; John McCormick, *The Global Environmental Movement*, 2d ed. (New York: Wiley, 1995), 27–45.

11. Ibid., 176–77. For similar comments from Eibl-Eibesfeldt about his 1954 trip to the Galápagos see Irenäus Eibl-Eibesfeldt, *Survey on the Galápagos Islands* (Paris: UNESCO, 1959), 8; and idem, "Galápagos: 'Enchanted Islands'," *UNESCO Courier*, 11

(January 1958): 20. For an ecologist's critique of his analysis see "Galápagos—the Noah's Ark of the Pacific," *Journal of Wildlife Management*, 27 (1963): 303–4.

12. Eibl-Eibesfeldt, *Galápagos*, 177.

13. Ibid., 21.

14. Eibl-Eibesfeldt, "Galápagos," 23.

15. Compare Julian Huxley, *Evolution: The Modern Synthesis* (New York: Harper, 1942), 326, with idem, *Evolution in Action* (New York: Harper, 1953), 70, and idem, "The Evolutionary Process," in Julian Huxley et al., eds., *Evolution as a Process* (London: George Allen, 1954), 6–7.

16. For example, Huxley called for a research station on the Galápagos "as a memorial to Darwin's great achievement." Julian Huxley, "The Galápagos Archipelago," *UNESCO Courier*, 11 (January 1958): 23.

17. For a critical analysis of this phase of Huxley's career see Michael Ruse, *Mystery of Mysteries: Is Evolution a Social Construction?* (Cambridge, MA: Harvard University Press, 1999), 96–99.

18. Daniel J. Kevles, "Huxley and the Popularization of Science," in C. Kenneth Waters and Albert Van Helden, eds., *Julian Huxley: Biologist and Statesman of Science* (Houston, TX: Rice University Press, 1992). With respect to Huxley's humanistic evolutionary ethics, Kevles adds, "Huxley here appeared to be deluding himself. The seemingly cosmic objectivity of his system was to a considerable extent shaped by [the primacy he placed on human personality]—and also by his resistance to that scourge of the twentieth century, totalitarianism." Ibid., 250.

19. Ruse, *Mystery of Mysteries*, 94. Huxley's earliest complete exposition of these views appeared in his aptly titled *Religion Without Revelation* (New York: Harper, 1927).

20. Julian Huxley, "Charles Darwin: Galápagos and After," in Robert I. Bowman, ed., *The Galápagos: Proceedings of the Symposia of the Galápagos International Scientific Project* (Berkeley: University of California Press, 1966), 8.

21. Julian Huxley, *Memories II* (London: George Allen, 1973), 13, 61, 249; idem, "Julian Huxley on his Mission at UNESCO," in Krishna R. Dronamraju, *If I Am to Be Remembered: The Life and Work of Julian Huxley with Selected Correspondence* (Singapore: World Scientific Publishing, 1993), 183; idem, "International Union for the Conservation of Nature (IUCN)," in ibid., 187; Victor A. Kovda, "Science," in United Nations Educational, Scientific and Cultural Organization, *In the Minds of Man: Unesco 1946 to 1971* (Paris: Unesco, 1972), 71. For a discussion of Huxley's role in promoting conservation as a mission within UNESCO and founding IUCN and the World Wildlife Fund see John

McCormack, *The Global Environmental Movement*, 2d ed. (Chichester, UK: John Wiley, 1995), 36–47. Linking these developments directly to the Galápagos, Huxley conceived of creating the IUCN during an official 1947 UNESCO visit to Ecuador, when his thoughts turned to the role that Galápagos wildlife played in convincing Darwin of the grand truth of evolution. "My recollection of this critical moment in Darwin's career was one of the factors that led me to propose the creation of IUCN," Huxley recalled twenty years later, "and this in turn persuaded Ecuador to make the archipelago a National Park—a Park which is also a natural biological laboratory, supported by the International Galápagos Foundation, where important studies on evolution and adaptation are now in progress." Huxley, *Memories II*, 41.

22. Julian Huxley, *UNESCO: Its Purpose and Philosophy* (Washington, DC: Public Affairs Press, 1947), 8, 39–40, 62.

23. Idem, "The Humanist Frame," in Julian Huxley, ed., *The Humanist Frame* (New York: Harper, 1961), 25.

24. Idem, *The Conservation of Wild Life and Natural Habitats in Central and East Africa* (Paris: Unesco, 1961), 60–91 (quote on 77). This report deals specifically with Huxley's proposal for national parks in East Africa but generally addresses the issue of wildlife parks in other such places. For further comments by Huxley on national parks see also idem, *The Human Crisis* (Seattle: University of Washington Press, 1963), 72–76.

25. "Such a view has important implications for science and education," Huxley added. "It implies that the most important sciences today . . . are evolutionary biology and ecology and their applications in scientific conservation." Huxley, *Human Crisis*, 23–24, 81–82. See also idem, "Humanist Frame," 43; idem, *Conservation of Wild Life*, 12.

26. Julian Huxley, "UNESCO, Paris, April 1959," in Dronamraju, *If I Am to Be Remembered*, 189. Huxley's comments did not relate expressly to conservation efforts for the Galápagos Islands but to those for similar wilderness areas, such as East African game parks.

27. Pacific Science Conference, "Proceedings," *Bulletin of the National Research Council*, 114 (1946): 43.

28. Mayr singles out six "main architects" of the neo-Darwinian synthesis: Huxley, Stebbins, Theodosius Dobzhansky, George Gaylord Simpson, Bernhard Rensch and himself. Ernst Mayr, *The Growth of Biological Thought: Diversity, Evolution, and Isolation* (Cambridge, MA: Harvard University Press, 1982), 568.

29. See, e.g., David Lack, "The Galápagos Finches (*Geospizinae*): A Study in Variation," *Occasional Papers of the California Academy of Sciences*, 21 (1945): 117, 135; idem, *Darwin's Finches* (Cambridge: Cambridge University Press, 1947), 26–30, 78–79, 91–95, 117, 134–48.

30. Robert I. Bowman, *Morphological Differentiation and Adaptation in the Galápagos Finches* (Berkeley: University of California Press, 1961), 1, 275; Robert I. Bowman, interview with author, 1 February 2000, Berkeley, CA.

31. Bowman, *Morphological Differentiation*, 262, 263, 268, 275.

32. Jonathan Weiner, *The Beak of the Finch: A Story of Evolution in Our Time* (New York: Vintage, 1995), 145–46. The split among biologists over the role of interspecific competition in the evolution of Darwin's finches is discussed in Ian Abbott et al., "Comparative Ecology of Galápagos Ground Finches (*Geospiza* Gould): Evaluation of the Importance of Floristic Diversity and Interspecific Competition," *Ecological Monographs*, 47 (1977): 153. For a sampling of the ongoing debate cf. Joseph H. Connell, "Diversity and the Coevolution of Competitors, or the Ghost of Competition Past," *Oikos*, 35 (1980): 131–38, and Daniel Simberloff, "The Great God of Competition," *The Sciences*, 24 (July–August 1984): 17–22, with the standard neo-Darwinian account in George Gaylord Simpson and William S. Beck, *Life: An Introduction to Biology*, 2d ed. (New York: Harcourt, Brace & World, 1965), 649.

33. Robert I. Bowman, *Report on a Biological Reconnaissance of the Galápagos Islands During 1957* (Paris: UNESCO, 1960), 7; idem, "The Scientific Need for Island Reserve Areas," Sixteenth International Congress of Zoology, *Proceedings*, 3 (Washington, DC: International Congress of Zoology, 1963), 394, 399; idem, "The Charles Darwin Foundation," *Pacific Discovery*, 24 (1971): 20–21; Bowman, interview with author, 1 February 2000.

34. Robert I. Bowman, "Academically Speaking," *Pacific Discovery*, 14 (September–October 1961): 37; idem, *Report on a Biological Reconnaissance*, 7, idem, interview with author, 1 February 2000; Eibl-Eibesfeldt, "Galápagos," 20. On Ripley's 1937 visit see Dillon Ripley, *Trail of the Monkey Bird: 30,000 Miles of Adventure with a Naturalist* (New York: Harper, 1940), 26–31.

35. Bowman, *Report on a Biological Reconnaissance*, 7–8; Eibl-Eibesfeldt, *Survey*, 8–9; Bowman, interview with author, 1 February 2000. According to Bowman's report to UNESCO, the expedition sought to study the possibilities for establishing a wildlife preserve and research station in the Galápagos, check the status of "vanishing" species and "obtain adequate photographic documentation of the islands and the biota for publicity purposes." Bowman, *Report on a Biological Reconnaissance*, 8.

36. Bowman, *Report on a Biological Reconnaissance*, 8; Eibl-Eibesfeldt, *Galápagos*, 8–9; Rudolf Freund and Alfred Eisenstaedt, "The Enchanted Isles," *Life*, 8 September 1958, 58–76; "A Showcase of Evolution," in *Evolution* (New York: Time-Life, 1962), 17–31. A photograph taken by Robert Bowman of a Galápagos marine iguana graced the cover of this first volume in the Life Nature Library.

37. Robert I. Bowman, "Treasure Islands of Science," *Américas*, December 1958, 19.

38. Eibl-Eibesfeldt, "Galápagos," 20.

39. Bowman, "Treasure Islands," 20.

40. Eibl-Eibesfeldt, *Survey*, 23–26; Bowman, *Report on a Biological Reconnaissance*, 59–60.

41. Bowman, "Treasure Islands," 21.

42. Eibl-Eibesfeldt, *Survey*, 27 (quote); Bowman, *Report on a Biological Reconnaissance*, 60.

43. See, e.g., Eibl-Eibesfeldt, *Survey*, 22; idem, "Challenge of Galápagos," 407.

44. Idem, *Survey*, 22–23.

45. Freund and Eisenstaedt, "Enchanted Isles," 59, 64–65, 76; See also "Showcase of Evolution," 25.

46. Huxley, "Galápagos Archipelago," 23.

47. Robert I. Bowman, "A Plea for an International Biological Station on the Galápagos Islands," and Harold J. Coolidge, "The Growth and Development of International Interest in Safeguarding Endangered Species," in the Fifteenth International Congress of Zoology, *Proceedings* (London: International Congress of Zoology, 1958), 58–59 (Coolidge here attributed this change among scientists to "the increase in our knowledge of animal ecology and a greater realization of the importance of preserving habitats").

48. Victor Van Straelen, Introduction, *Occasional Papers of the California Academy of Sciences*, 44 (1962): 7; Jean Dorst and Jacques Laruelle, *The First Seven Years of the Charles Darwin Foundation for the Galápagos Isles, 1959–1966* (Brussels: Charles Darwin Foundation, 1967), 5–7, 23–27; Roger Perry, *The Galápagos Islands* (New York: Dodd, Mead, 1972), 85–87; Peter Scott, *Charles Darwin Foundation for the Galápagos Islands* (Brussels: Charles Darwin Foundation, 1961), 9; "IUCN," *IUCN Bulletin*, 9 (1960): 1. The other two initial foundation officers were Secretary General Jean Dorst of Paris and Vice President Luis Jaramillo of Paris and Quito.

49. Julian Huxley, "Discussion," in Bowman, "Plea for Biological Station," 59. Ripley here added, "Publicity by itself could be harmful, and must be correlated with action."

50. Bryan Nelson, *Galápagos: Islands of Birds* (New York: William Morrow, 1968), xv, 4.

51. Julian Huxley to Armand Denis, 2 June 1960, and Julian Huxley to Rene Maheu, 27 April, 1965, in Dronamraju, *If I Am to Be Remembered*, 194, 205 (both letters address conservation efforts in Africa). For a general discussion of the promotion of wildlife films by Huxley and Osborn see Gregg Mitman, *Reel Nature: America's Romance with Wildlife on Film* (Cambridge, MA: Harvard University Press, 1999), 195.

52. Mitman, *Reel Nature*, 129–30.

53. Jack C. Couffer, "Galápagos Adventure," *Natural History*, 65 (1956): 141, 144, 145 (Disney filmmaker's quote); Peter Scott, "BBC/IUCN Darwin Centenary Expedition," in The Wildfowl Trust, *11th Annual Report* (1958–1959), 73; Peter Scott and Philippa Scott, *Faraway LOOK Two* (London: Cassell, 1962), 116 (Scott quote); Scott, *Charles Darwin Foundation*, 4; Sven Gillsäter, *From Nature to Nature: Oases of the Animal World in the Western Hemisphere* (London: Allen & Unwin, 1968), 103 (Gillsäter quote).

54. Christian Zuber, *Animal Paradise* (New York: Barnes, 1966), 75.

55. Jean Dorst, "Where Time Stood Still: The Galápagos Islands and Their Prehistoric Creatures," *UNESCO Courier*, 14 (September 1961): 31.

56. Irenäus Eibl-Eibesfeldt, "A Revisit of the Galápagos," *Noticias de Galápagos*, 5–6 (1965): 24.

57. Zuber, *Animal Paradise*, 96.

58. Gillsäter, *From Nature to Nature*, 86–87.

59. Jerold M. Lowenstein, "Galápagos Journal," *California Monthly*, July–August 1965, 14–15; Nathan Cohen and Robert Bowman, "The Galápagos International Scientific Project," *Noticias de Galápagos*, 3 (1964): 5; Bowman, interview with author, 1 February 2000.

60. For extended analysis of these developments see, e.g., Daniel J. Kevles, *The Physicists: The History of a Scientific Community in Modern America* (Cambridge, MA: Harvard University Press, 1971), 287–366; Daniel S. Greenberg, *The Politics of Pure Science: An Inquiry into the Relationship Between Science and Government in the United States* (New York: Times Mirror, 1967), 51–179.

61. House Subcommittee on Science, Research, and Development, Committee on Science and Astronautics, *The National Science Foundation: A General Review of Its First 15 Years*, 89th Cong., 2d sess., 24 January 1966, 32; Senate Special Committee on Atomic Energy, *Atomic Energy Hearings*, 79th Cong., 2d sess., 3 December 1945, 170 (Russell quote); J. Merton England, *A Patron for Pure Science: The National Science Foundation's Formative Years, 1945–57* (Washington, DC: National Science Foundation, 1982), 83–179. By 1963, annual federal spending for science exceeded $12 billion (or triple the total spent by the federal government for science during the Second World War), with about $1 billion earmarked for biological research. By that time, one NSF report noted, "the Federal Government initiates and finances most of the research and development in the Nation." National Science Foundation, *Federal Funds for Science XI: Fiscal Years 1961, 1962, and 1963* (Washington, DC: Government Printing Office, 1963), 2–7, 23–25, 29–31 (quote on 4).

62. "Galápagos Symposium: Notes on Meeting of July 12, 1962," Robert T. Orr Papers, box 77, California Academy of Sciences (CAS) Archives, San Francisco, CA. For the NSF's mandate see National Science Foundation Act of 1950, Public Law 507, 81st Cong., 2d sess., 10 May 1950; England, *Patron for Pure Science*, 227–310.

63. In addition to naming Huxley, the application stated that scientists of "the caliber of Theodosius Dobzhansky, Father Guillermo Kuschel, David Lack, Robert Mertens, Robert Cushman Murphy, Bernhard Rensch, and Rafael Rodriguez are expected to participate." University of California, Berkeley, "Galápagos International Scientific Project, October 1962" (proposal to the National Science Foundation), 8–9, Robert T. Orr Papers, box 77, CAS Archives.

64. "In Darwin's Footsteps," *Newsweek*, 20 January 1964, 52.

65. David Perlman, "The Blowout for Iguanas," *San Francisco Chronicle*, 23 January 1964, 1, 16 (quote); idem, "Galápagos Quest Is On," *San Francisco Chronicle*, 11 January 1964, 1, 6; Lowenstein, "Galápagos Journal," 7–12; Robert I. Bowman, "The Galápagos International Scientific Project," *Book of Knowledge, 1965* (New York: Grolier, 1965), 359; Cohen and Bowman, "Galápagos International Scientific Project," 4–11; Bowman, interview with author, 1 February 2000. From the early planning stages, organizers intended to invite political figures who could influence conservation decisions; for example, a 1963 letter from Orr to a potential major donor noted about the dedication ceremonies, "It is planned to try to bring the President of Ecuador, whoever he will be then." Robert T. Orr to Kenneth K. Bechtel, 9 August 1963, Robert T. Orr Papers, box 77, CAS Archives. At the time, Ecuador's elected president had been recently overthrown by a military coup, whose leaders still held power when the dedication occurred.

66. "Final Report on *The Galápagos International Scientific Project*," 17–19, Robert T. Orr Papers, box 77, CAS Archives. By arousing international pressure and domestic support for protecting the Galápagos, the report here noted that "the Galápagos Project provided an appropriate time and proper public-opinion environment for final action by the Ecuadorian Government" on a practical plan for island conservation.

67. Carl B. Koford, "Economic Resources of the Galápagos Islands," in Bowman, ed., *The Galápagos*, 290. Huxley stated about the Galápagos Project in his opening address, "The overriding aim must be with the general idea of conservation and special conservation areas, to make people understand that the conservation of wild life and natural beauty is of great value and importance to the human species, and outstandingly so on the Galápagos." Huxley, "Galápagos and After," ibid., 9.

68. David Perlman, "Hunt for 250,000 Bugs," *San Francisco Chronicle*, 17 February 1964, 1; Ira L. Wiggins and Duncan M. Porter, *Flora of the Galápagos Islands* (Stanford, CA:

Stanford University Press, 1971), vii–viii, 43–44; Bowman, interview with author, 1 February 2000. Regarding this expedition to the Galápagos Islands, an early planning document stated, "While it was agreed that few major research projects could be completed in one or two months, it was felt that such an undertaking with a number of prominent scientists would . . . focus world attention, as well as that of the Government of Ecuador, on . . . the necessity of conserving their most valuable biological assets." Alan E. Leviton, "Report to the Belvedere Scientific Fund for a Grant-in-Aid to Cover the Expenses of Participation of Members of the Staff of the California Academy of Sciences in the Galápagos Islands International Scientific Project," 1, in Robert T. Orr Papers, box 77, CAS Archives.

69. Allyn G. Smith, "Our Man on the Galápagos," *Pacific Telephone Magazine*, September 1964, 32 (identifies the effort "to protect endemic Galápagos birds and animals" as "a main purpose of our expedition"). Lowenstein, "Galápagos Journal," 13; David Perlman, "Camp Was Never Like This," *San Francisco Chronicle*, 23 February 1964, 17; Bowman, "Galápagos Project" (Grolier), 359; "Galápagos Goiter Hunter," *San Francisco Chronicle*, 20 February 1964, 8; "Final Report," 18; Paul A. Colinvaux, "A Galápagos Symposium," *Ecology*, 48 (1967): 702 (quote); D. W. Snow, "Research Station in the Galápagos," *Oryx*, 7 (1964): 275.

70. Eibl-Eibesfeldt, "Revisit of the Galápagos," 24–25.

Chapter 8

1. Loren Eiseley, "Rock Redondo: Science and Literature in the Galápagos," in Kenneth Brower, ed., *Galápagos: The Flow of Wildness*, vol. 1 (San Francisco: Sierra Club, 1968), 26–27.

2. Idem, *Darwin's Century: Evolution and the Men Who Discovered It* (Garden City, NY: Doubleday, 1958), 152, 172–74.

3. Idem, "Rock Redondo," 32.

4. Ruth Moore and the Editors of *Life*, *Evolution* (New York: Time Incorporated, 1962), 31.

5. Charles Darwin, *On the Origin of Species*, facs. 1st ed. (Cambridge: Cambridge University Press, 1964), 402.

6. Among the many leading high-school and college biology textbooks from the 1960s that feature such accounts see Helene Curtis, *Biology* (New York: Natural History Press, 1968), 699–702, 744–45; William T. Keeton, *Biological Science* (New York: Norton, 1967), 691–95; Willis H. Johnson et al., *Principles of Zoology*, 2d ed. (New York: Holt, 1977), 610–11; Biological Sciences Curriculum Study, *High School Biology: BSCS Green Version*

(Chicago: Rand McNally, 1963), 573–75; George Gaylord Simpson and William S. Beck, *Life: An Introduction to Biology*, 2d ed. (New York: Harcourt, 1965), 457–60. For representative examples of pre-Lack textbooks that do not mention Galápagos finches see Perry D. Strausbaugh and Bernard R. Wiemer, *General Biology: A Textbook for College Students* (New York: Wiley, 1938), 488 (briefly notes Darwin's visit to the Galápagos); Alfred C. Kinsey, *An Introduction to Biology* (Philadelphia: Lippincott, 1926), 4 (briefly mentions Galápagos Tortoises).

7. Edward O. Wilson, *The Diversity of Life* (New York: Norton, 1992), 101–2.

8. Moore et al., *Evolution*, 18.

9. Eiseley, "Rock Redondo," 39–40.

10. Robert I. Bowman, "Contributions to Science from the Galápagos," in Roger Perry, ed., *Galápagos: Key Environments* (Oxford: Pergamon Press, 1984), 278.

11. For a sample of such geological analysis see Tom Simkin, "Geology of Galápagos Islands," in Perry, ed., *Galápagos*, 16–38; Allan Cox, "Ages of the Galápagos Islands," in Robert I. Bowman et al., eds., *Patterns of Evolution in Galápagos Organisms* (San Francisco: AAAS, 1983), 11–24; D. M. Christie et al., "Drowned Islands Downstream from the Galápagos HotSpot Imply Extended Speciation Time," *Nature*, 355 (1992): 246–48; Reinhard Werner et al., "Drowned 14-m.y.-old Galápagos Archipelago Off the Coast of Costa Rica: Implications for Tectonic and Evolutionary Models," *Geology*, 27 (1999): 499–502.

12. Darwin, *Origin of Species*, 388–406. See generally David Quammen, *The Song of the Dodo: Island Biogeography in an Age of Extinctions* (New York: Scribner, 1996), 115–258. These factors relating to the study of evolution on islands were consistently mentioned by scientists interviewed for this book: included are P. Dee Boersma (7 March 2000), Robert I. Bowman (1 February 2000) Duncan M. Porter (11–12 August 2000), Dolph Schluter (16 March 2000) and Howard L. Snell (20 January 2000). Bowman's views on the biological significance of islands also appear in Robert I. Bowman, "The Nature of Islands," *Pacific Discovery*, 48 (Summer 1995): 8–17.

13. Peter R. Grant, "Patterns on Islands and Microevolution," in Peter R. Grant, ed., *Evolution on Islands* (Oxford: Oxford University Press, 1998), 1.

14. See Heidi M. Snell et al., *Galápagos Bibliography, 1535–1995* (Quito, Ecuador: Charles Darwin Foundation, 1996), which lists over 7,500 Galápagos science articles, with most of them dating from the last twenty-five years covered by the listing.

15. Theodosius Dobzhansky et al., *Evolution* (San Francisco: Freeman, 1977), 186. From his historic work in the western South Pacific, Mayr stresses that evolution may proceed differently on different islands, which he sees as the wonder of biological laws: their di-

versity in contrast with the laws of physics. Ernst Mayr, interview with author, 22 December 2000, Cambridge, MA.

16. Duncan M. Porter, "Vascular Plants of the Galápagos: Origins and Dispersal," in Bowman et al., eds., *Patterns of Evolution*, 36 (quote); idem, "Endemism and Evolution in Terrestrial Plants, in Perry, ed., *Galápagos*, 90–99; idem, "Relationships of the Galápagos Flora," *Biological Journal of the Linnean Society*, 21 (1984): 243–49.

17. E. Yale Dawson, "The Giants of Galápagos," *Natural History*, 71 (November 1962): 53–55 (quote on 55); Porter, "Vascular Plants," 40–43.

18. Deborah Clark, "Native Land Mammals," in Perry, ed., *Galápagos*, 225–26. Clark obviously did not include humans in this particular contest.

19. James L. Patton and Mark S. Hafner, "Biosystematics of the Native Rodents of the Galápagos Archipelago, Ecuador," in Bowman et al., eds., *Patterns of Evolution*, 540–62; Robert T. Orr, "Evolutionary Aspects of the Mammalian Fauna of the Galápagos," in Robert I. Bowman, ed., *The Galápagos: Proceedings of the Symposia of the Galápagos International Scientific Project* (Berkeley: University of California Press, 1966), 276–80 (Orr added that Galápagos bats are also smaller than mainland types). Sheer size is not a certain predictor of evolutionary fitness, of course, but among mammals of the same or similar species, it is related to strength and survival ability in gross terms.

20. Quammen, *Song of the Dodo*, 157.

21. MacFarland served as the station's resident director from 1974 to 1978, followed by several terms as president of the Charles Darwin Foundation. Snell continues to live and work at the station much of each year.

22. Craig MacFarland, "Goliaths of the Galápagos," *National Geographic Magazine*, 131 (1967): 539–45 (quote on 540). See also Thomas H. Fritts, "Morphometrics of Galápagos Tortoises: Evolutionary Implications," in Bowman et al., eds., *Patterns of Evolution*, 115–21; James L. Patton, "Genetical Processes in the Galápagos," *Biological Journal of the Linnean Society*, 21 (1984): 102.

23. Kornelia Rassmann, "Evolutionary Age of the Galápagos Iguanas Predates the Present Galápagos Islands," *Molecular Phylogenetics and Evolution*, 7 (1997): 158–72; Jeff S. Wyles and Vincent M. Sarah, "Are the Galápagos Iguanas Older than the Galápagos? Molecular Evolution and Colonization Models for the Archipelago," in Bowman et al., eds., *Patterns of Evolution*, 180–83.

24. Howard L. Snell et al., "Variation Among Populations of Galápagos Land Iguana (*Conolophus*): Contrasts of Phylogeny and Ecology," *Biological Journal of the Linnean Society*, 21 (1985): 185–202; Howard L. Snell, interview with author, 15 January 2000, Puerto Ayora, Galápagos; Irenäus Eibl-Eibesfeldt, "The Large Iguanas of the Galápagos Is-

lands," in Perry, ed., *Galápagos*, 157–65; P. Dee Boersma, "An Ecological Study of the Galápagos Marine Iguana," in Bowman et al., eds., *Patterns of Evolution*, 161–73; Martin Wikelski and Fritz Trillmich, "Body Size and Sexual Size Dimorphism in Marine Iguanas Fluctuate as a Result of Opposing Natural and Sexual Selection: An Island Comparison," *Evolution*, 51 (1997): 930–34; Jacques-Yves Cousteau and Philippe Diolé, *Three Adventures: Galápagos, Titicaca, The Blue Hole* (Garden City, NY: Doubleday, 1973), 17, 41–58, 67–76 (quote on 17).

25. MacFarland, "Goliaths," 640; Cousteau and Diolé, *Three Adventures*, 55, 77; Snell, interview with author.

26. Charles Darwin, *Voyage of the Beagle* (London: Penguin, 1989), 288.

27. Bryan Nelson, *Galápagos: Islands of Birds* (New York: Morrow, 1968), xvii, 319 (quotes); idem, "The Man-o'-War Bird," *Natural History*, 75 (May 1966): 33–39; Robert I. Bowman, "Contributions to Science from the Galápagos," in Perry, ed., *Galápagos*, 303.

28. Jonathan Weiner, "The Handy-Dandy Evolution Prover," *New York Times Magazine*, 8 May 1994, 41.

29. Idem, *Beak of the Finch: A Story of Evolution in Our Time* (New York: Vintage, 1995), 9.

30. Peter R. Grant, "Hybridization of Darwin's Finches on Isla Daphne Major, Galápagos," *Philosophical Transactions of the Royal Society of London B*, 340 (1993): 127; idem, "The Role of Interspecific Competition in the Adaptive Radiation of Darwin's Finches," in Bowman et al., eds., *Patterns of Evolution*, 195.

31. Idem, "Convergent and Divergent Character Displacement," *Biological Journal of the Linnean Society*, 4 (1972): 63–64; idem, "The Classical Case of Character Displacement," *Evolutionary Biology*, 8 (1975): 329–30; idem, *Ecology and Evolution of Darwin's Finches* (Princeton, NJ: Princeton University Press, 1986), xii.

32. Compare the neo-Darwinian view of Darwin's finches as set forth in A. J. Nicholson, "The Role of Population Dynamics in Natural Selection," in Sol Tax, ed., *The Evolution of Life: Its Origin, History and Future* (Chicago: University of Chicago Press, 1960), 500–501; and Ernst Mayr, *Animal Species and Evolution* (Cambridge, MA: Harvard University Press, 1963), 83–84; with the classic presentation of the environmentalist view in H. G. Andrewartha and L. C. Birch, *The Distribution and Abundance of Animals* (Chicago: University of Chicago Press, 1954), 20–26, advanced during the early 1970s in, e.g., J. Reddingius and P. J. Den Boer, "Simulation Experiments Illustrating Stabilization of Animal Numbers by Spreading of Risk," *Oecologia*, 5 (1970): 240–84. For further criticism of Lack's interpretation see Dennis Chitty, "What Regulates Bird Populations?" *Ecology*, 48 (1967): 698–701.

33. Grant, *Ecology and Evolution*, xii; Ian Abbott et al., "Comparative Ecology of Galápagos Ground Finches (*Geospiza* Gould): Evaluation of the Importance of Floristic Diversity and Interspecific Competition," *Ecological Monographs*, 47 (1977): 153; Peter R. Grant to author, 6 December 2000, in author's files.

34. Peter R. Grant and B. Rosemary Grant, "Speciation and Hybridization of Birds on Islands," in Peter R. Grant, ed., *Evolution on Islands* (Oxford: Oxford University Press, 1998), 142; Dolph Schluter, interview with author, 16 March 2000, Vancouver, British Columbia.

35. W. L. Brown, Jr., and Edward O. Wilson, "Character Displacement," *Systematic Zoology*, 5 (1956): 52.

36. Mayr, *Animal Species*, 83; Darwin, *Origin of Species*, 111.

37. Grant, *Ecology and Evolution*, xii; idem, "Classical Case," 173 (the "classical case" in this title refers to rock nuthatches).

38. Abbott et al., "Comparative Ecology," 153.

39. Grant, *Ecology and Evolution*, xii; Ian Abbott et al., "Seed Selection and Handling Ability of Four Species of Darwin's Finches," *The Condor*, 77 (1975): 332–33; Peter R. Grant et al., "Darwin's Finches: Population Variation and Natural Selection," *Proceedings of the National Academy of Sciences*, 73 (1976): 257–58.

40. Grant, "Role of Interspecific Competition," 187.

41. Abbott et al., "Comparative Ecology," 159, 167; see also Abbott et al., "Seed Selection," 332–35. For Lack's continuing resistance to environmentalist interpretations of Darwin's finches see David Lack, "Subspecies and Sympatry in Darwin's Finches," *Evolution*, 23 (1969): 252–53.

42. Grant, "Role of Interspecific Competition," 190–91; see also Abbott et al., "Comparative Ecology," 157–77; Peter R. Grant, "Speciation and the Adaptive Radiation of Darwin's Finches," *American Scientist*, 69 (1981): 656–60.

43. For Grant's willingness to challenge Lack's ideas, see, e.g., Peter R. Grant, "Island Birds," *Bird Banding*, 48 (1977): 299 (Grant's review of Lack's last book).

44. Grant, "Role of Interspecific Competition," 195 (emphasis added).

45. Idem, *Ecology and Evolution*, xiii.

46. Idem, "Role of Interspecific Competition," 195.

47. Ibid.

48. Peter T. Boag and Peter R. Grant, "Intense Natural Selection in a Population of Darwin's Finches (*Geospininae*) in the Galápagos," *Science*, 214 (1981): 82–84. See also Peter T. Boag, "The Biology of Darwin's Finches on Daphne Major Island," *Charles Darwin Research Station Annual Report, 1978* (typescript, Charles Darwin Research Station li-

brary, Puerto Ayora, Galápagos), 4; Grant, *Ecology and Evolution*, 180–94; idem, "The Endemic Land Birds," in Perry, ed., *Galápagos*, 182; Peter T. Boag and Peter R. Grant, "Heritability of External Morphology in Darwin's Finches," *Nature*, 274 (1978): 794; Peter T. Boag, "The Heritability of External Morphology in Darwin's Ground Finches (*Geospiza*) on Isla Daphne Major, Galápagos," *Evolution*, 37 (1983): 877–94.

49. Wiener, *Beak of the Finch*, 81; idem, "Handy-Dandy Evolution Prover," 40. See also B. Rosemary Grant and Peter R. Grant, "Evolution of Darwin's Finches Caused by a Rare Climatic Event," *Proceedings of the Royal Society of London B*, 251 (1993): 114.

50. H. Lisle Gibbs and Peter R. Grant, "Oscillating Selection on Darwin's Finches," *Nature*, 327 (1987): 511–12. See also Grant and Grant, "Evolution of Darwin's Finches," 111–14.

51. Grant and Grant, "Evolution of Darwin's Finches," 111.

52. National Academy of Sciences, *Teaching About Evolution and the Nature of Science* (Washington, DC: National Academy Press, 1998), 19. The National Academy of Sciences here paraphrased Grant's prediction that in the original read: "If droughts occur once a decade, on average, repeated directional selection at this rate with no selection in between droughts would transform one species into another within 200 years." Peter R. Grant, "Natural Selection and Darwin's Finches," *Scientific American*, 273 (October 1991): 86.

53. Weiner, "Handy-Dandy Evolution Prover," 40.

54. Phillip E. Johnson, Op-Ed, *Wall Street Journal*, 16 August 1999, A14. See also Jonathan Wells, *Icons of Evolution: Science or Myth?* (Washington, DC: Regnery, 2000), 168–70: "It remains a theoretical possibility that the various species of Galápagos finches originated through natural selection. But the Grants' observations provided no direct evidence for this." Ibid., 170.

55. See, e.g., B. Rosemary Grant, "Selection on Bill Characters in a Population of Darwin's Finches: *Geospiza conirostris* on Isla Genovesa, Galápagos," *Evolution*, 39 (1985): 523–32.

56. Weiner, *Beak of the Finch*, 19.

57. See, e.g., Peter R. Grant and B. Rosemary Grant, "Hybridization of Bird Species," *Science*, 256 (1992): 193–97; Grant, "Hybridization of Darwin's Finches," 127.

58. Kenneth Petren et al., "A Phylogeny of Darwin's Finches Based on Microsatellite DNA Length Variation," *Proceedings of the Royal Society of London B*, 266 (1999): 321, 327–28.

59. Daniel Simberloff and Edward F. Connor, "Missing Species Combinations," *American Naturalist*, 118 (1981): 236. See also Edward F. Connor and Daniel Simberloff, "Species Number and Compositional Similarity of the Galápagos Flora and Avifauna," *Ecological Monographs*, 48 (1978): 230–36; Donald R. Strong, Jr., et al., "Tests of Community-Wide Character Displacement Against Null Hypotheses," *Evolution*, 33 (1979): 899–911.

60. See, e.g., John A. Hendrickson, Jr., "Community Wide Character Displacement Reexamined," *Evolution*, 35 (1981): 801; Rauno V. Alatalo, "Bird Species Distributions in the Galápagos and Other Archipelagos: Competition or Chance?" *Ecology*, 63 (1982): 882–85. Peter Grant and Ian Abbott vigorously defended their conclusions in Peter R. Grant and Ian Abbott, "Interspecific Competition, Island Biogeography and Null Hypotheses," *Evolution*, 34 (1980): 332–41.

61. Michael E. Gilpin and J. M. Diamond, "Are Species Co-occurrences on Islands Nonrandom, and Are Null Hypotheses Useful in Community Ecology?" in Donald R. Strong, Jr., et al., eds., *Ecological Communities: Conceptual Issues and the Evidence* (Princeton, NJ: Princeton University Press, 1984), 315. This volume contains the tepid concession by Simberloff, "So we appear to agree that Galápagos ground finches are not random subsets, that interspecific competition is likely at least partly responsible for this, and that the strongest evidence resides in bill features of coexisting races and species." Daniel Simberloff, "Morphological and Taxonomic Similarity and Combinations of Coexisting Birds in Two Archipelagos," ibid., 252.

62. Dolph Schluter and Peter R. Grant, "The Distribution of *Geospiza difficilis* in Relation to *G. fuliginosa* in the Galápagos Islands: Tests of Three Hypotheses," *Evolution*, 36 (1982): 1221–25; idem, "Ecological Correlates of Morphological Evolution in a Darwin's Finch, *Geospiza difficilis*," *Evolution*, 38 (1984): 865–68; Dolph Schluter, "Determinants of Morphological Patterns in Communities of Darwin's Finches," *American Naturalist*, 123 (1984): 190–94; Dolph Schluter and Peter R. Grant, "Ecological Character Displacement in Darwin's Finches," *Science*, 227 (1985): 1056–58.

63. Schluter, interview with author, 16 March 2000; Weiner, *Beak of the Finch*, 151.

64. Peter R. Grant, "Recent Research on the Evolution of Land Birds on the Galápagos," *Biological Journal of the Linnean Society*, 21 (1984): 132. The biological community generally supports Grant's view; see, e.g., Wilson, *Diversity of Life*, 176.

65. Robert I. Bowman, "Contributions to Science from the Galápagos," in Perry, ed., *Galápagos*, 305 (quote); Robert I. Bowman, "The Evolution of Song in Darwin's Finches," in Bowman et al., eds., *Patterns of Evolution*, 318–24; Robert I. Bowman, "Ecology and Evolution," *Wilson Bulletin*, 100 (1988): 160–63 (book review); Robert I. Bowman, interview with author, 1 February 2000, Berkeley, CA. Regarding the proposal for Genovesa ground finches see Joseph Vagvolgyi and Maria W. Vagvolgyi, "The Taxonomic Status of the Small Ground-Finch, *Geospiza* of Genovesa Island, Galápagos, and Its Relevance to Interspecific Competition," *Auk*, 106 (1989): 144–47. For Grant's response see Peter R. Grant et al., "The Allopatric Phase of Speciation: The Sharp-Beaked Ground

Finch (*Geospiza difficilis*) on the Galápagos Islands," *Biological Journal of the Linnean Society*, 69 (2000): 313, in which the authors used phylogenetic, morphologic and behavioral evidence to conclude that the Genovesa flock constitutes a "well-differentiated population" radiating from the sharp-beak ground finch species. "This does not undermine the competition argument but it does negate the Bowman/Vagvolgyi challenge," Grant notes. Grant to author.

66. Schluter, interview with author; Grant, *Ecology and Evolution*, 404 (quote). See also Weiner, *Beak of the Finch*, 9–10, 19, 57–58.

67. See, e.g., Bowman, "Contributions," 305.

68. Grant, *Ecology and Evolution*, 405.

69. These scientific debates, involving what the participants describe as "high stakes at the high table" of evolutionary theory, are highly public. See, e.g., Niles Eldredge, *Reinventing Darwin: The Great Debate at the High Table of Evolutionary Theory* (New York: John Wiley & Sons, 1995), 125. Participants have written scores of popular books in addition to countless scholarly articles. The best known of the partisans defending gradualism and pervasive adaptionism at the individual or gene level include Richard Dawkins, John Maynard Smith, William Hamilton, George Williams and Daniel Dennett. The best known of the partisans questioning the sufficiency of the neo-Darwinian synthesis to account wholly for the evolutionary process include Steven Jay Gould, Niles Eldredge and Richard Lewontin. My review of their books found little mention of Darwin's finches or other Galápagos species except to establish the "fact" of evolution.

70. See, e.g., J. W. Klotz, "Flora and Fauna of the Galápagos Islands," *Creation Research Society Quarterly*, 9 (1972): 15–16 (accepts evolutionists' taxonomy), 17 (first quote); Walter E. Lamberts, "The Galápagos Island Finches," *Creation Research Society Quarterly*, 3 (1966): 76–77 ("It seems much more in line with reality to consider these birds as all one species, broken up into various island forms"); Duane T. Gish, "A Decade of Creationist Research," *Creation Research Society Quarterly*, 12 (1975): 37–38, 41–42 (second quote); Phillip Johnson, *Darwin on Trial* (Washington, DC: Regnery, 1991), 25–27.

71. Nelson Montoya, interview with author, 18 January 2000, Puerto Ayora, Galápagos (Loma Linda teacher); Robert Bensted-Smith, interview with author, 19 January 2000, Puerto Ayora, Galápagos (director, Charles Darwin Research Station).

72. David Hull, "God of the Galápagos," *Nature*, 352 (1991): 485.

73. David Denby, "In Darwin's Wake," *New Yorker*, 21 July 1997, 56, 61.

74. Ibid., 50 (quote), 61 (frigate bird experience).

75. Roger Tory Peterson, "The Galápagos: Eerie Cradle of New Species," *National Geographic Magazine*, 142 (April 1967): 541, 585 (quote).

76. Compare Herman Melville, "The Encantadas, or Enchanted Isles," in Harrison Hayford et al., eds., *The Piazza Tales and Other Prose Pieces: 1839–1860* (Evanston, IL: Northwestern University Press, 1987), 128–29, and Charles Darwin to J. D. Hooker [8 or 15 July 1846], in Frederick H. Burkhardt et al., eds., *The Correspondence of Charles Darwin*, vol. 3 (Cambridge: Cambridge University Press, 1985), 327.

77. Weiner, *Beak of the Finch*, 82.

78. For a sample of public opinion on this issue in the United States see "Believer Nation," *Public Perspective*, 11 (May–June 2000): 35.

79. E. C. Agassiz, "A Cruise Through the Galápagos," *Atlantic Monthly*, 31 (May 1973): 580, 582 (quote).

80. David L. Lack, "My Life as an Amateur Ornithologist," *Ibis*, 115 (1973): 429.

81. Ernst Mayr and Alister C. Hardy quoted in "In Appreciation," *Ibis*, 115 (1973): 432–34, 436.

82. David Lack, *Evolutionary Theory and Christian Belief: The Unresolved Conflict* (London: Methuen, 1961), 72, 77, 111–14. For Gould's description of the concept, which Lack fits better than most examples cited by Gould, see Stephen Jay Gould, *Rocks of Ages: Science and Religion in the Fullness of Life* (New York: Ballantine, 1999).

83. Charles Darwin to Asa Gray, 22 May 1860, in Frederick H. Burkhardt et al., eds., *The Correspondence of Charles Darwin*, vol. 8 (Cambridge: Cambridge University Press, 1993), 224. For a discussion of this issue see Edward J. Larson and Larry Witham, "Scientists and Religion in America," *Scientific American*, 281 (September 1999): 88–93.

84. Ernst Mayr, *This Is Biology: The Science of the Living World* (Cambridge, MA: Harvard University Press, 1997), 33.

85. John C. Avise, *The Genetic Gods: Evolution and Belief in Human Affairs* (Cambridge, MA: Harvard University Press, 1998), 208–9.

Chapter 9

1. Annie Dillard, "Innocence in the Galápagos," *Harper's Magazine*, 250 (May 1975): 74, 77, 78.

2. Idem, *Pilgrim at Tinker Creek* (New York: Harper's Magazine Press, 1974), 5–9, 65 (creation quote), 175–76 (evolution quote).

3. Linda L. Smith, *Annie Dillard* (New York: Twayne, 1991), 10. Dillard received the 1975 Front Page Award for Excellence from the New York Newswomen's Club for her Galápagos article. She received the Pulitzer Prize in general nonfiction that same year for *Pilgrim at Tinker Creek*.

4. Dillard, "Innocence in the Galápagos," 77.

5. Ibid., 79–82. In the only change that she made to this article when published in a book of her essays, Dillard added a paragraph extending her criticism of Darwinism and its application to humans. See Annie Dillard, *Teaching a Stone to Talk: Expeditions and Encounters* (New York: Harper, 1982), 120.

6. See, e.g., Margaret Loewen Reimer, "The Dialectical Vision of Annie Dillard's *Pilgrim at Tinker Creek*," *Critique*, 24 (1983): 182–83; Smith, *Annie Dillard*, 44, 125.

7. Dillard, "Innocence in the Galápagos," 82.

8. Irving Johnson and Electra Johnson, "The *Yankee*'s Wander-world," *National Geographic Magazine*, 95 (1949): 8.

9. Idem, "Lost World of the Galápagos," *National Geographic Magazine*, 115 (1959): 691.

10. Irving Johnson, "A Cruising Guide to the Galápagos Islands," *Yachting*, 99 (March 1956): 58, 60, 116.

11. Frank Rohr, "We Sailed to the Galápagos," *Travel*, 108 (December 1957): 49–50. For similar accounts of island life during the 1950s from UNESCO investigators see Robert I. Bowman, "Treasure Islands of Science," *Américas* (December 1958): 21; Irenäus Eibl-Eibesfeldt, *Survey on the Galápagos Islands* (Paris: UNESCO, 1959), 22; idem, "Challenge of the Galápagos," *Nature Magazine*, 50 (1957): 407.

12. Johnson, "Cruising Guide," 61; Nanci Badger, "Exploring the Galápagos," *Yachting*, 126 (October 1969): 66; James A. W. Crowe, "Fascinating Wildlife Remains Only Lure to Primitive Galápagos," *Detroit News*, 16 June 1969; Carleton Mitchell, "Cruise to Islands of Wonders and Terror," *Sports Illustrated*, 8 June 1970, 40.

13. Eliot Porter, "Galápagos Journal," in Kenneth Brower, ed., *Galápagos: The Flow of Wildness*, vol. 2 (San Francisco: Sierra Club, 1968), 39, Compañia Ecuatoriana de Turismo, "Visit the Galápagos" (Guayaquil, 1964), in Robert T. Orr Papers, box 77, California Academy of Sciences Archives, San Francisco, CA; Badger, "Exploring the Galápagos," 65.

14. Nelson Doubleday and C. Earl Cooley, *Encyclopedia of World Travel*, rev. ed., vol. 1 (Garden City, NJ: Doubleday, 1967), 597.

15. Carl B. Koford, "Economic Resources of the Galápagos Islands," in Robert I. Bowman, ed., *The Galápagos: Proceedings of the Symposia of the Galápagos International Scientific Project* (Berkeley: University of California Press, 1966), 290.

16. I. R. Grimwood and D. W. Snow, "Recommendations on the Administration of the Proposed Galápagos National Park and the Development of Its Tourist Potential," June 1966, 1, Vertical File 333.783, "Parque Nacional," Charles Darwin Research Station Library, Puerto Ayora, Galápagos Islands (hereafter Darwin Station Library).

17. Ibid., 11. So long as their approach was followed, the planners here concluded, "It is our firm belief that the scenic and wild life attractions of the Galápagos Islands are such that the proposed national park could become one of the most popular of its kind, capable of attracting a steady flow of visitors from all over the world." See also G. T. Corley Smith, "A Brief History of the Charles Darwin Foundation for the Galápagos Islands, 1959–1988," *Noticias de Galápagos*, 49 (1990): 15.

18. H. N. Hoeck, "Socio-Economic Development in Galápagos: Consequences for an Unique Island Ecosystem," in W. Erdelen et al., eds., *Tropical Ecosystems: Proceedings of the International and Interdisciplinary Symposium* (Weikersheim: Margarf Scientific Books, 1991), 164–69; Smith, "Brief History," 15–16; Diana Argentina Costa, "Ecuadorian and Non-Ecuadorian Visitors to the Galápagos Islands, Ecuador: A Comparison," (master's thesis, University of Idaho, 1991), 8; Suzanne Wiedel, "Galápagos," *Travel & Camera*, 33 (June 1970): 89; James A. O. Crowe, "Galápagos Casts a Spell," *Detroit News*, 20 June 1969, C11.

19. Arthur D. Little, Inc., *Tourism in the Galápagos Islands: An Introductory Brochure* (Cambridge, MA: Arthur D. Little, 1967), 7–8.

20. Ibid., 7.

21. Richard Atcheson, "Galápagos—The Way the World Was," *Holiday*, 48 (September 1971): 73.

22. Paul J. C. Friedlander, "Galápagos: A Return to Life's Earliest Beginnings," *New York Times*, 15 February 1950, sec. 10, 3.

23. Ila Stanger, "The Galápagos: Irreplaceable Islands," *Harper's Bazaar*, 105 (November 1971): 64.

24. Atcheson, "Galápagos," 53, 103.

25. Ibid., 53.

26. Walter S. Wingo, "A Fragile Land Where Man Is Greatest Enemy," *U.S. News & World Report*, 27 April 1981, 70. For figures see Costa, "Ecuadorian and Non-Ecuadorian Visitors," 8–9; Hoeck, "Socio-Economic Development," 164–66; "Galápagos Science Conference, 1972," *Noticias de Galápagos*, 22 (1974): 15; "Island Population Figures," at Darwin Station Library.

27. Atcheson, "Galápagos," 89.

28. "A Ten Year Projection of Activities of the Charles Darwin Foundation," *Noticias de Galápagos*, 17 (1971): 19–20.

29. "The Master Plan," *Noticias de Galápagos*, 23 (1975): 6–12; Tui De Roy Moore, "Effects of Tourism: Observations of a Resident Naturalist," *Noticias de Galápagos*, 32 (1980): 12. The plan drew directly on UNESCO's Man and the Biosphere initiative, which repre-

sented a recent development in designing nature preserves. Under the UNESCO scheme, an ideal biosphere reserve would contain concentric zones with varying degrees of restriction on human use. For a description of the Man and the Biosphere initiative see William P. Gregg, Jr., and Betsy Ann McGean, "Biosphere Reserves: Their History and Their Promise," *Orion Nature Quarterly*, 4 (1985): 42–88.

30. "Ten Year Projection," 19–21; "Galápagos Science Conference," 17, 19; G. T. Corley Smith, "Looking Back on Twenty Years of the Charles Darwin Foundation," *Noticias de Galápagos*, 30 (1979): 12; David Davies, "Kill a Pinta Goat a Day," *Nature*, 249 (1974): 788–89; Kim Heacox, "Preserving the Enchanted Isles," *Américas*, 36 (November 1984): 47–49.

31. See, e.g., Ralph Hubley, "Saving the Galápagos," *Christian Science Monitor*, 28 November 1972, 9; Wingo, "Fragile Land," 69; Guy Mountfort, "The Problems of Tourism to Island Reserves," *Noticias de Galápagos*, 15–16 (1970): 12; Moore, "Effects of Tourism," 12–13.

32. Atcheson, "Galápagos," 53.

33. Irenäus Eibl-Eibesfeldt, "Twenty Years After," *Noticias de Galápagos*, 24 (1976): 5.

34. John Tierney, "Lonesome George of the Galápagos," *Science 85*, 6 (June 1985): 55. See also "Giant Tortoise Breeding at the Darwin Station," *Noticias de Galápagos*, 22 (1974): 20; G. T. Corley Smith, "Looking Back," *Noticias de Galápagos*, 45 (1987): 13; idem, "Brief History," 14–15, 20.

35. Thomas H. Fritts and Patricia R. Fritts, eds., *Race with Extinction: Herpetological Notes of J. R. Slevin's Journey to the Galápagos, 1905–1906* (Albuquerque, NM: Herpetologists' League, 1982), 83.

36. Tierney, "Lonesome George," 50, 59.

37. Bruce Wallace, "A Treat to Wildlife," *MaClean's*, 15 April 1985, 50; Cruz Marquez, "The Giant Tortoises and the Great Fire on Isabela," *Noticias de Galápagos*, 44 (1986): 8.

38. Smith, "Looking Back," 10 (quote), 13.

39. S. Dillon Ripley, "The View from the Castle," *Smithsonian*, 9 (December 1978): 8. See generally Cruz Marquez et al., "The Ten Year Struggle to Save the Endangered Land Iguanas," *Noticias de Galápagos*, 44 (1986): 9; Frederic Golden, "Visit to the Enchanted Isles," *Time*, 26 June 1978, 71; Robert Bersted-Smith, interview with author, 19 January 2000, Puerto Ayora, Galápagos.

40. Smith, "Brief History," 21, 24, 27, 32; Peter Kramer, "Galápagos Conservation: Present Position and Future Outlook," *Noticias de Galápagos*, 22 (1974): 4; "Control of Introduced Animals," *Noticias de Galápagos*, 31 (1980): 4; Robin Nelson, "The Galápagos: Ecology's Prize Dilemma," *Popular Mechanics*, 150 (1978): 98–99, 206–7; Wingo, "Fragile Land," 69–70; Duncan M. Porter, interview with author, 12 August 2000, Cambridge, UK. In

his theory of natural selection, Darwin explained why fitter introduced species often displace less-fit native ones on oceanic islands. In the Galápagos, however, the Darwin Station sides with the less-fit native forms in an effort to preserve primordial conditions for scientists and tourists alike. "What hope for Man's future, to live in harmony with such a laboratory of nature's creation," Ripley wrote in 1977 about the station's successes. "The world is better for this unity of purpose in an essential cause." S. Dillon Ripley, "The View from the Castle," *Smithsonian*, 8 (1977): 8. In 2000, the Darwin Station reported success in its long battle to eradicate goats from Pinta Island. "Achievements of a Busy Year," *Galápagos News*, Summer 2000, 2.

41. David Finkelstein and Grace D. Polk, "Preserving the World's Heritage: Galápagos National Park," *Oceans*, 15 (May 1982): 67. Tourist numbers include only registered visitors to the national park—more certainly entered without registering. Population figures for the archipelago vary greatly: figures cited herein use an approximation derived from the varying estimates. For figures see Hoeck, "Socio-Economic Development," 164–67; Costa, "Ecuadorian and Non-Ecuadorian Visitors," 6–10; Fundación Natura and World Wide Fund for Nature, *Galápagos Report, 1997–1998* (Quito, Ecuador: Fundación Natura, 1998), 26–27, 30–31, 44–45; Jerry Emory, "Managing Another Galápagos Species—Man," *National Geographic Magazine*, 173 (January 1988): 147; Fritz Trillmich, "Conservation Problems on the Galápagos: The Showcase of Evolution in Danger," *Naturwissenschaften*, 79 (1992): 4.

42. "That it exists at all comes as a surprise to many tourists," the Darwin Station visitor guide observed in 1988 about the town. "Today a stroll through downtown Puerto Ayora at the peak of the tourist season is unforgettable. Motorcycles, bicycles, and patchwork cars kick up dust on the main road . . . . Hotels, restaurants, and houses are under construction everywhere. It soon becomes apparent that big changes are afoot in this frontier-like town." Emory, "Managing Another Galápagos Species," 147, 149.

43. Miguel Cifuentes, quoted in Golden, "Visit to the Enchanted Isles," 71; idem, quoted in Emory, "Managing Another Galápagos Species," 150.

44. Emory, "Managing Another Galápagos Species," 150 (quote). See also Martha Honey, "Paying the Price of Ecotourism," *Américas*, 46 (November–December 1994): 47; Smith, "Brief History," 33; Michael D'Orso, interview with author, 19 January 2000, Puerto Ayora, Galápagos; Godfrey Merlin, interview with author, 20 January 2000, Puerto Ayora, Galápagos. By 1996, four out of five Galápagos residents lived in urban areas. Fundación Natura and World Wide Fund for Nature, *Galápagos Report, 1996–1997* (Quito, Ecuador: Fundación Natura, 1997), 1 (also notes the high rate of immigration from the mainland). For a comparison of economic conditions on the Ecuadorian

mainland and the Galápagos and for data on and analysis of social conditions (including the number of schools and public opinion on the Darwin Station and island conservation) see Fundación Natura, *Galápagos Report, 1997–1998*, 32–33, 39–43, 77. The Darwin Station continues to promote community education by training local teachers, publishing teaching materials and producing broadcast materials. See, e.g., "Achievement of a Busy Year," 3.

45. Sid Kane, "Extinction of the Species?" *World Development* (United Nations Development Programme), 4 (November 1991): 15.

46. Peter Benchley, "Galápagos: Paradise in Peril," *National Geographic Magazine*, 195 (1999): 16, 28 (quote); Richard Stone, "Fishermen Threaten Galápagos," *Science*, 267 (1995): 612; Tad Friend, "The Mild Kingdom," *Vogue*, 185 (September 1995): 394; Fundación Natura, *Galápagos Report, 1997–1998*, 30–38.

47. Dan Duane, "Where the Girls Are," *Surfer*, 38 (August 1997): 111, 113, 116. For a typical discussion of traveling to the Galápagos for recreational scuba diving see Marty Snyderman, "Enchanted Islands of the Extremes," *Skin Diver*, 48 (1999): 86–87.

48. Myers Robertson, "Isla Bartolomé, Galápagos," *Sports Illustrated*, 20 February 1998, 28 (quote on fold-out page), 160–74; Laurence Shames, "Evolution's Lab—While It Lasts," *Business Week*, 1 April 1996, 88–89.

49. For representative news reports on the episode see Michael D. Lemonick, "Can the Galápagos Survive?" *Time*, 30 October 1995, 80–82; Shames, "Evolution's Lab," 89; Stone, "Fishermen Threaten Galápagos," 611–12; Friend, "Mild Kingdom," 394.

50. Harry Thurston, "Last Look at Paradise?" *International Wildlife*, 27 (May–June 1997): 18. See generally Theodore Macdonald, *Los Conflictos en las Islas Galápagos: Analisis y Recommendacions para so Manejo* (Quito, Ecuador: Fundación Charles Darwin, 1997), 10–18.

51. Benchley, "Galápagos," 16.

52. Friend, "Mild Kingdom," 398.

53. Benchley, "Galápagos," 1–6, 29.

54. Trillmich, "Conservation Problems," 1 (Trillmich quote), 4–5; Snell quoted in Public Broadcasting System, "Voyage to the Galápagos," transcript of Show 1001 (1999), pt. 2, 10.

55. Public Broadcasting System, "Voyage to Galápagos," pt. 2, 10 (Causton quote); Honey, "Paying the Price," 47 (Izurieta and Fritts quotes); Fundación Natura, *Galápagos Report, 1997–1998*, 54–57; Benchley, "Galápagos," 30; Howard Snell, interview with author, 20 January 2000, Puerto Ayora, Galápagos.

56. See, e.g., Smith, "Brief History," 14–33; Washington Tapia Aguilera, "40 Años Dedicados a la Conservación," *Parque Nacional Galápagos*, 2 (1999): 6, 11, 18; Tierney, "Lonesome George," 50–59; Bersted-Smith, interview with author, 19 January 2000.

57. David A. Westbrook, "The New Regime for Conservation in the Galápagos: Ensuring Effectiveness and Compliance," address to 80th Annual Meeting of the American Association for the Advancement of Science—Pacific Division, 22 June 1999, 3–4, 8; Alison Abbott, "Conservationists Seek Compromise over New Threat to Galápagos," *Nature*, 383 (1996): 3; "New Law to Protect Galápagos," *Science*, 279 (1998): 1857. In this address, Westbrook concluded that "the way to achieve the conservation goals embodied in the New Law is continually and objectively to monitor conservation problems in the Galápagos, and to ensure that such information is internationally available."

58. Letter from Heidi Snell to author, 21 November 2000, in author's file; Tui De Roy, "Circular Messages," 15–28 November 2000, in author's files; Fundación Natura et al., "Pressure by a Group of Fishermen Puts the Participative Process in Jeopardy," 21 November 2000 (press release, in author's files); Larry Rohter, "Where Darwin Mused, Strife Over Ecosystem," *New York Times*, 27 December 2000, A1, A8 (fisherman's quote); Environment News Service, "Galápagos Lobster Fishermen Smash Up Conservation Offices," 22 November 2000 (news release, in author's files).

59. Smith, "Looking Back," 14; "Estación Cientifica Charles Darwin, Programa de Cientificos Visitantes," Charles Darwin Research Station, Puerto Ayora, Ecuador, 1999; Cooperacion Externa (Charles Darwin Research Station) to author, 9 March 2000, in author's files. Ninety student interns volunteered to work on conservation projects at the Darwin Station in 1999, including sixty from Ecuador.

60. Leigh Newman, "You Say You Want an Evolution?" *Travel Holiday*, 182 (January 2000): 94. Countering with the promise of deep-sea diving among 20-foot hammerhead sharks, three resort-class scuba centers compete with tour boat operators for tourist dollars in present-day Puerto Ayora. Enterprising young settlers offer surfing classes at the local beach.

61. Charles Darwin to J. D. Hooker, [8 or 15 July 1846], in Frederick H. Burkhardt et al., eds., *The Correspondence of Charles Darwin*, vol. 3 (Cambridge: Cambridge University Press, 1985), 327.

62. John F. Ross, "IMAX Takes Us Undersea in the Galápagos," *Smithsonian*, 30 (October 1999): 55, 62.

63. John B. Corliss et al., "Submarine Thermal Springs on the Galápagos Rift," *Science*, 203 (1979): 1073–82; J. Frederick Grassle, "Hydrothermal Vent Animals: Distribution and Biology," *Science*, 229 (1985): 713–17; Holger W. Jannasch and Michael J. Mottl, "Geomicrobiology of Deep-Sea Hydrothermal Vents," *Science*, 229 (1985): 717–25.

64. Peter R. Grant and B. Rosemary Grant, "Hybridization of Bird Species," *Science*, 256 (1992): 196 (quote); Peter R. Grant, "Hybridization of Darwin's Finches on Isla Daphne

Major, Galápagos," *Philosophical Transactions of the Royal Society of London B*, 340 (1993): 138; Peter R. Grant and B. Rosemary Grant, "Phenotypic and Genetic Effects of Hybridization in Darwin's Finches," *Evolution*, 48 (1994): 310–13; Ernst Mayr, interview with author, 20 April 2000, Cambridge, MA. Evolution theory critic Jonathan Wells goes further to assert that the finding of hybridization among Galápagos finches undermines their historic status as "icons" of Darwinism. "The increase in average beak size in several species of Galápagos finches after a severe drought—and its return to normal after the drought ended—is direct evidence for natural selection in the wild," he writes. "As examples of the origin of species by natural selection, however, Darwin's finches leave a lot to be desired." Jonathan Wells, *Icons of Evolution: Science or Myth?* (Washington, DC: Regnery, 2000), 173. The reference to a hybrid or "heterogeneous swarm" appeared in Percy R. Lowe, "The Finches of the Galápagos in Relation to Darwin's Conception of Species," *Ibis*, 78 (1936): 311.

65. See, e.g., David J. Anderson, "Evolution of Obligate Siblicide in Boobies: 1. A Test of the Insurance-Egg Hypothesis," *American Naturalist*, 135 (1990): 339–42; idem, "Evolution of Obligate Siblicide in Boobies: 2. Food Limitation and Parent-Offspring Conflict," *Evolution*, 44 (1990): 2072–76; Public Broadcasting System, "Voyage to Galápagos," pt. 2, 6 (quote).

66. Stephen Jay Gould, "Galápagos Islands," *National Geographic Traveler*, 16 (October 1999): 98–100.

67. Tom Stoppard, "Paradise and Purgatory," *Observer Magazine*, 29 November 1981, 40.

68. Idem, "This Other Eden," *Noticias de Galápagos*, 34 (1981): 6.

69. Herman Melville, "The Encantadas, or Enchanted Isles," in Harrison Hayford et al., eds., *The Piazza Tales and Other Prose Pieces: 1839–1860* (Evanston, IL: Northwestern University Press, 1987), 130.

70. Ursala Goodenough, *The Sacred Depths of Nature* (New York: Oxford University Press, 1998), 174; David Lack, *Evolutionary Theory and Christian Belief: The Unresolved Conflict* (London: Methuen, 1961), 36–37 (quote), 72–79 (evolutionary life through individual death).

INDEX

Abbott, Ian, 206–9
Abbott, Lynette, 208
Abingdon Island, 55, 93, 105; *see also* Pinta
 Island
Aca, 156
Academy, 132–38, 145
Academy Bay; development of, 175, 184,
 196, 228, 233; research at, 137, 168,
 182–87, 197, 224; *see also* Puerto
 Ayora
Acosta, José de, 18–20, 23, 33
Acushnet, 4
Adams, C. F., 127
adaptive radiation, 171–72, 198–202,
 214–15, 240
Agassiz, Alexander, 107–09, 114, 123
Agassiz, Elizabeth Cary, 98–100, 150, 218,
 265n30
Agassiz, Louis, 95–105, 117, 122–25, 145,
 218–19, 241
Albatross, 107–9, 267n52
Albmarle Island, 5, 9, 151; Darwin on, 68,
 70, 74; exploration of, 38, 42–49, 55,
 57; settlement of, 134, 138; tortoises
 on, 105–6, 117, 135–38; *see also* Isabela
 Island
Aldabra Island (Indian Ocean), 103,
 116
Alvin, 240
Amazon River, 96–97, 148
Amblyrhynchus cristatus, 202; *see also*
 iguana, marine
Ambrose, St., 16
America, early scientific views of, 16–20,
 33–34

American Museum of Natural History
 (New York), 102, 118, 146–56, 168
Anaximander, 15
Anderson, David J., 240–41
Angermeyer brothers, 224
Animal Species and Evolution (Mayr),
 207
Arcturus, 153–55
Arcturus Adventure (Beebe), 154–55
Aristotle, 15, 19
Astor, Vincent, 153, 156–57
Atcheson, Richard, 227–28
Atlantic Monthly, 96–100, 150, 218
Audubon, 227
Autobiography of Charles Darwin, 65
Avise, John C. 219–20

Bacon, Francis, 27, 34, 37
Baconian science, 27–30, 36–37
Baker, Joseph, 38
Baird, Spencer, 108–111
Baldwin, Carole, 239
Baltra Island, 175–77, 181, 185, 190,
 200–1, 204, 225–26
Baur, George, 111–116, 126–27, 142, 152,
 164–65, 272n25
Beagle, 8, 56–78, 85
Beak of the Finch (Weiner), xi, 184, 217
Beebe, William, 149–55, 157–58, 170,
 194, 279n24
Beck, Rollo, 120, 128–50, 219, 275n55
Bell, Thomas, 78
bestiaries, 19
Berlanga, Fray Tomás de, 21–23, 241
Bindloe Island, 93, 117, 135

311